FALLACIES IN MATHEMATICS

FALLACIES IN MATHEMATICS

BY

E. A. MAXWELL, Ph.D.

Fellow of Queens' College, Cambridge

CAMBRIDGE
AT THE UNIVERSITY PRESS
1969

PUBLISHED BY
THE SYNDICS OF THE CAMBRIDGE UNIVERSITY PRESS
Bentley House, 200 Euston Road, London, N.W.1
American Branch: 32 East 57th Street, New York, N.Y. 10022

Standard Book Number: 521 05700 0

First printed 1959
Reprinted 1961
First paperback edition 1963
Reprinted 1969

First printed in Great Britain at the University Press, Cambridge
Reprinted by photolithography in Great Britain
by Bookprint Limited, Crawley, Sussex

CONTENTS

PREFACE

The aim of this book is to instruct through entertainment. The general theory is that a wrong idea may often be exposed more convincingly by following it to its absurd conclusion than by merely denouncing the error and starting again. Thus a number of by-ways appear which, it is hoped, may amuse the professional and help to tempt back to the subject those who thought they were losing interest.

The standard of knowledge expected is quite elementary; anyone who has studied a little deductive geometry, algebra, trigonometry and calculus for a few years should be able to follow most of the exposition with no trouble.

Several of the fallacies are well known, though I have usually included these only when I felt that there was something fresh to add. There is not (I hope) much of the rather outworn type

$$x = 0,$$

$$\therefore\ x(x-1) = 0,$$

$$\therefore\ x-1 = 0,$$

$$\therefore\ x = 1,$$

$$\therefore\ 1 = 0.$$

I have also tried to avoid a bright style; the reader should enjoy these things in his own way.

My original idea was to give references to the sources of the fallacies, but I felt, on reflection, that this was to give them more weight than they could carry. I should, however, thank the editor of the *Mathematical Gazette* for his ready permission to use many examples which first appeared there.

It was with pleasure that I received the approval of the Council of the Mathematical Association to arrange for the Association to receive one half of the royalties from the sale of this book. I welcome the opportunity to record my gratitude for much that I have learned and for many friendships that I have made through the Association.

I must express my thanks to members (past and present) of the staff of the Cambridge University Press, who combined their skill and care with an encouragement which, in technical jargon, became real and positive. I am also indebted for valuable advice from those who read the manuscript on behalf of the Press, and to my son, who helped me to keep the proofs from becoming unnecessarily fallacious.

E. A. MAXWELL

CAMBRIDGE
28 April 1958

CHAPTER I

THE MISTAKE, THE HOWLER AND THE FALLACY

All mathematicians are wrong at times. In most cases the error is simply a MISTAKE, of little significance and, one hopes, of even less consequence. Its cause may be a momentary aberration, a slip in writing, or the misreading of earlier work. For present purposes it has no interest and will be ignored.

The HOWLER in mathematics is not easy to describe, but the term may be used to denote an error which leads *innocently* to a *correct* result. By contrast, the FALLACY leads by *guile* to a *wrong* but plausible conclusion.

One or two simple examples will illustrate the use of these terms. The first is a howler of a kind that many people, here and abroad, might think the British system of units only too well deserves:

To make out a bill:

$\frac{1}{4}$ lb. butter	@ 2*s*. 10*d*. per lb.
$2\frac{1}{2}$ lb. lard	@ 10*d*. per lb.
3 lb. sugar	@ $3\frac{1}{4}$*d*. per lb.
6 boxes matches	@ 7*d*. per dozen.
4 packets soap-flakes	@ $2\frac{1}{2}$*d*. per packet.

(Who but an examiner would include four packets of soap-flakes in such an order?)

The solution is

$$8\tfrac{1}{2}d. + 2s.\ 1d. + 9\tfrac{3}{4}d. + 3\tfrac{1}{2}d. + 10d. = 4s.\ 8\tfrac{3}{4}d.$$

One boy, however, avoided the detailed calculations and simply added all the prices on the right:

$$2s.\ 10d. + 10d. + 3\tfrac{1}{4}d. + 7d. + 2\tfrac{1}{2}d. = 4s.\ 8\tfrac{3}{4}d.$$

The innocence of the pupil and the startling accuracy of the answer raise the calculation to the status of howler.

The emphasis of this book is on the fallacy, though a selection of howlers (leading, sometimes, to curious generalisations) will be given in the last chapter. Two synthetic howlers may, however, be added here before we proceed to the real business:

(i) *To prove the formula*

$$a^2 - b^2 = (a-b)(a+b).$$

Consider the quotient

$$\frac{a^2 - b^2}{a - b}.$$

'Cancel' an a and a b:

$$\frac{\not{a}^{\not{2}} - \not{b}^{\not{2}}}{\not{a} - \not{b}},$$

and then 'cancel' the minus 'into' the minus:

$$\overset{+}{\frac{\not{a}^{\not{2}} \not{-} \not{b}^{\not{2}}}{\not{a} \not{-} \not{b}}}.$$

The result is $a + b$.

(ii) *To simplify* $\frac{26}{65}$ *and* $\frac{16}{64}$.

The correct answers are obtained by cancelling:

$$\frac{2\not{6}}{\not{6}5} = \frac{2}{5}$$

and

$$\frac{1\not{6}}{\not{6}4} = \frac{1}{4}.$$

Consider next a typical fallacy:

To prove that every triangle is isosceles.

Let ABC (Fig. 1) be a given triangle. It is required to prove that AB is *necessarily* equal to AC.

If the internal bisector of the angle A meets BC in D, then, by the angle-bisector theorem,

$$\frac{DB}{AB} = \frac{DC}{AC}.$$

Now

$$\angle ADB = \angle ACD + \angle CAD$$
$$= C + \tfrac{1}{2}A,$$

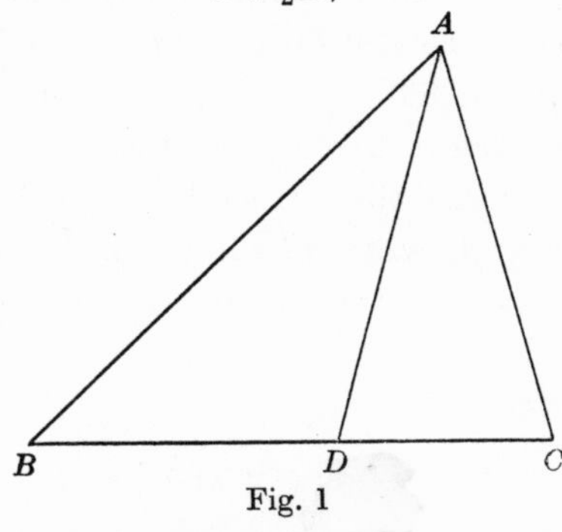

Fig. 1

so that, by the sine rule applied to the triangle ADB,

$$\frac{DB}{AB} = \frac{\sin BAD}{\sin ADB}$$
$$= \frac{\sin \frac{1}{2}A}{\sin (C + \frac{1}{2}A)}.$$

Further,

$$\angle ADC = \angle ABD + \angle BAD$$
$$= B + \tfrac{1}{2}A,$$

so that

$$\frac{DC}{AC} = \frac{\sin \frac{1}{2}A}{\sin (B + \frac{1}{2}A)}.$$

Hence

$$\frac{\sin \frac{1}{2}A}{\sin (C + \frac{1}{2}A)} = \frac{\sin \frac{1}{2}A}{\sin (B + \frac{1}{2}A)}.$$

Moreover $\sin \frac{1}{2}A$ is not zero, since the angle A is not zero, and so

$$\sin (C + \tfrac{1}{2}A) = \sin (B + \tfrac{1}{2}A),$$

or

$$C + \tfrac{1}{2}A = B + \tfrac{1}{2}A,$$

or

$$C = B.$$

The triangle is therefore isosceles.

The analysis of this fallacy may be used to illustrate standard features which we shall meet often.

To begin with, of course, the actual error must be detected. (The false step here follows the line

$$\sin(C+\tfrac{1}{2}A) = \sin(B+\tfrac{1}{2}A).$$

Equality of sine need not mean equality of angle.) But this, in a good fallacy, is only a part of the interest, and the lesser part at that.

The second stage is to effect, as it were, a *reconciliation statement* in which (i) the correct deduction is substituted for the false and, when possible, (ii) the discrepancy between the wrong and the correct theorems is accounted for in full.

Thus the step

$$\sin(C+\tfrac{1}{2}A) = \sin(B+\tfrac{1}{2}A)$$

leads not only to the stated conclusion

$$C+\tfrac{1}{2}A = B+\tfrac{1}{2}A,$$

but also to the alternative

$$C+\tfrac{1}{2}A = 180^\circ-(B+\tfrac{1}{2}A),$$

or
$$A+B+C = 180^\circ.$$

The necessity for the angles B, C to be equal is negatived by the fact that the sum of the angles A, B, C is *always* 180°. The error is therefore found and the correct version substituted.

In the work that follows we shall usually begin with a straightforward statement of the fallacious argument, following it with an exposure in which the error is traced to its most elementary source. It will be found that this process may lead to an analysis of unexpected depth, particularly in the first geometrical fallacy given in the next chapter.

CHAPTER II

FOUR GEOMETRICAL FALLACIES ENUNCIATED

There are four well-known fallacies whose statement requires only the familiar theorems of elementary (schoolboy) geometry. They are propounded without comment in this chapter, the analysis, which becomes very fundamental, being postponed. The reader will doubtless wish to detect the errors for himself before proceeding to their examination in the next chapter.

¶ 1. THE FALLACY OF THE ISOSCELES TRIANGLE.

To prove that every triangle is isosceles.

GIVEN: A triangle ABC (Fig. 2).

REQUIRED: To prove that, necessarily,

$$AB = AC.$$

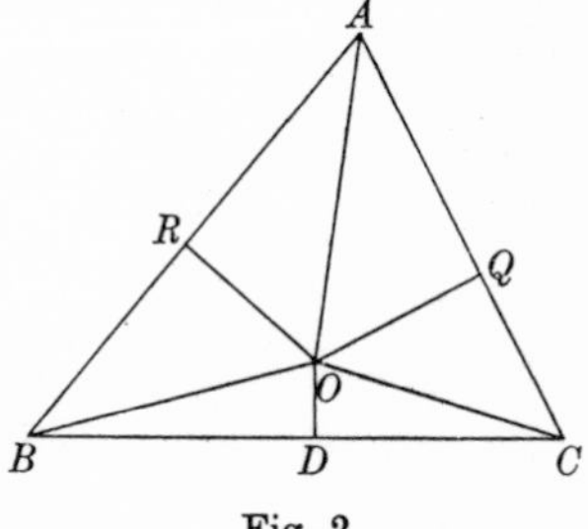

Fig. 2

CONSTRUCTION: Let the internal bisector of the angle A meet the perpendicular bisector of BC at O. Draw OD, OQ, OR perpendicular to BC, CA, AB respectively.

PROOF: Since

$$DO = DO$$
$$DB = DC$$
$$\angle ODB = \angle ODC$$
$$\therefore \triangle ODB \equiv \triangle ODC \qquad \text{(SAS)*}$$
$$\therefore OB = OC.$$

* Notation such as SAS is used as an abbreviation for 'having two sides and the included angle equal'. The symbol $\equiv$ denotes congruence.

Also $$AO = AO,$$
$$\angle RAO = \angle QAO$$
$$\angle ARO = \angle AQO$$
$$\therefore \triangle ARO \equiv \triangle AQO \qquad \text{(ASA)}$$
$$\therefore AR = AQ$$
and $$OR = OQ.$$

Hence, in triangles OBR, OCQ,
$$\angle ORB = \angle OQC = \text{right angle},$$
$$OB = OC \qquad \text{(proved)}$$
$$OR = OQ \qquad \text{(proved)}$$
$$\therefore \triangle ORB \equiv \triangle OQC \qquad \text{(rt. } \angle\text{, H, S)}$$
$$\therefore RB = QC.$$

Finally, $$AB = AR + RB$$
$$= AQ + QC \qquad \text{(proved)}$$
$$= AC. \qquad \text{Q.E.D.}$$

¶ 2. THE FALLACY OF THE RIGHT ANGLE.

To prove that every angle is a right angle.

GIVEN: A square $ABCD$ and a line BE drawn outwards from the square so that $\angle ABE$ has a given value, assumed obtuse (Fig. 3). (If the value is acute, take $\angle ABE$ to be its supplement.)

REQUIRED: To prove that
$$\angle ABE = \text{a right angle.}$$

CONSTRUCTION: Let P, Q be the middle points of CD, AB. Take BE equal in length to a side of the square, and let the perpendicular bisector of DE meet PQ, produced if necessary, at O.

PROOF: By symmetry, PQ is the perpendicular bisector of CD and of AB.

Consider the triangles ORD, ORE:

$$OR = OR$$

$$\angle ORD = \angle ORE \qquad \text{(construction)}$$

$$RD = RE \qquad \text{(construction)}$$

$$\therefore \triangle ORD \equiv \triangle ORE. \qquad \text{(SAS)}$$

In particular,

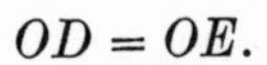

$$OD = OE.$$

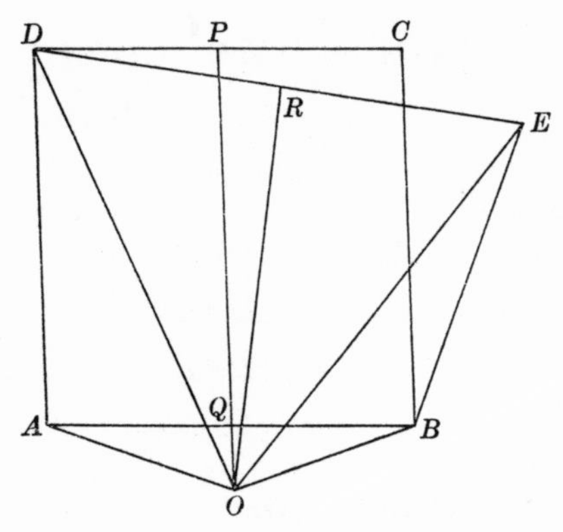

Fig. 3

Consider the triangles OQA, OQB:

$$OQ = OQ$$

$$\angle OQA = \angle OQB \qquad \text{(perpendicular bisector)}$$

$$QA = QB \qquad \text{(construction)}$$

$$\therefore \triangle OQA \equiv \triangle OQB. \qquad \text{(SAS)}$$

In particular, $OA = OB$

and (for later reference)

$$\angle OAB = \angle OBA.$$

Consider the triangles OAD, OBE:

$$OA = OB \quad \text{(proved)}$$

$$AD = BE \quad (BE = \text{side of square})$$

$$OD = OE \quad \text{(proved)}$$

$$\therefore \triangle OAD \equiv \triangle OBE. \quad \text{(SSS)}$$

In particular,

$$\angle OAD = \angle OBE.$$

Now the point O on PQ may be either (i) between P, Q; (ii) at Q; (iii) beyond Q (as in Fig. 3).

In case (i)

$$\begin{aligned} \angle ABE &= \angle OBE + \angle OBA \\ &= \angle OAD + \angle OAB \quad \text{(proved)} \\ &= \text{right angle.} \end{aligned}$$

In case (ii)

$$\begin{aligned} \angle ABE &\equiv \angle OBE \\ &= \angle OAD \quad \text{(proved)} \\ &\equiv \angle BAD \\ &= \text{right angle.} \end{aligned}$$

In case (iii)

$$\begin{aligned} \angle ABE &= \angle OBE - \angle OBA \\ &= \angle OAD - \angle OAB \quad \text{(proved)} \\ &= \text{right angle.} \end{aligned}$$

Hence, in all three cases,

$$\angle ABE = \text{one right angle.}$$

❡ 3. The Trapezium Fallacy.

To prove that, if ABCD is a quadrilateral in which $AB = CD$, then AD is necessarily parallel to BC.

GIVEN: A quadrilateral $ABCD$ (Fig. 4) in which $AB = CD$.

REQUIRED: To prove that AD is parallel to BC.

CONSTRUCTION: Draw the perpendicular bisectors of AD, BC. If they are parallel, the theorem is proved. If not, let them meet in O, which may be outside (Fig. 4 (i)) or inside (Fig. 4 (ii)) the quadrilateral.

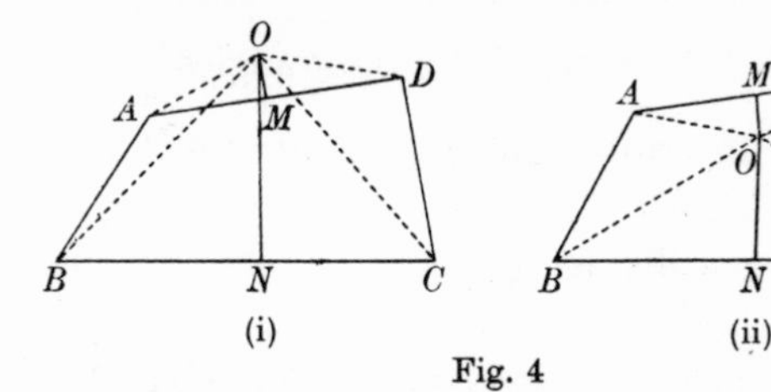

Fig. 4

PROOF: The perpendicular bisector of a line is the locus of points equally distant from its ends. Hence

$$OA = OD$$

$$OB = OC.$$

Compare the triangles OAB, ODC.

$$OA = OD \qquad \text{(proved)}$$

$$OB = OC \qquad \text{(proved)}$$

$$AB = DC \qquad \text{(given)}$$

$$\therefore \ \triangle OAB \equiv \triangle ODC \qquad \text{(SSS)}$$

$$\therefore \ \angle OAB = \angle ODC.$$

Also, by isosceles triangles,

$$\angle OAD = \angle ODA. \qquad (OA = OD)$$

Hence, by subtraction (Fig. 4.(i)) or addition (Fig. 4 (ii)),

$$\angle BAD = \angle CDA.$$

Similarly

$$\angle OBA = \angle OCD \quad (\triangle OAB \equiv \triangle ODC)$$

and

$$\angle OBC = \angle OCB. \qquad (OB = OC)$$

Hence, by addition,

$$\angle ABC = \angle DCB.$$

Thus

$$\angle BAD + \angle ABC = \angle CDA + \angle DCB.$$

But the sum of these four angles is four right angles, since they are the angles of the quadrilateral $ABCD$. Hence

$$\angle BAD + \angle ABC = 2 \text{ right angles.}$$

Also these are interior angles for the lines AD, BC with transversal AB.

Thus AD is parallel to BC.

¶ 4. THE FALLACY OF THE EMPTY CIRCLE.

To prove that every point inside a circle lies on its circumference.

GIVEN: A circle of centre O and radius r, and an arbitrary point P inside it (Fig. 5).

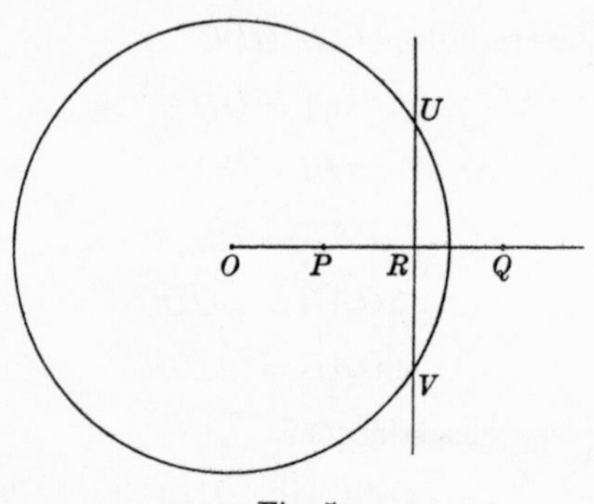

Fig. 5

REQUIRED: To prove that P lies on the circumference.

CONSTRUCTION: Let Q be the point on OP produced beyond P such that

$$OP.OQ = r^2,$$

and let the perpendicular bisector of PQ cut the circle at U, V. Denote by R the middle point of PQ.

PROOF:

$$
\begin{aligned}
OP &= OR - RP \\
OQ &= OR + RQ \\
&= OR + RP \qquad (RQ = RP,\ \text{construction}) \\
\therefore\ OP.OQ &= (OR - RP)(OR + RP) \\
&= OR^2 - RP^2 \\
&= (OU^2 - RU^2) - (PU^2 - RU^2) \quad \text{(Pythagoras)} \\
&= OU^2 - PU^2 \\
&= OP.OQ - PU^2 \qquad (OP.OQ = r^2 = OU^2) \\
\therefore\ PU &= 0
\end{aligned}
$$

$\therefore$ P is at U, on the circumference.

CHAPTER III

DIGRESSION ON ELEMENTARY GEOMETRY

Elementary geometry is commonly taught in the schools for two main purposes: to instil a knowledge of the geometrical figures (triangle, rectangle, circle) met in common experience, and also to develop their properties by logical argument proceeding step by step from the most primitive conceptions. The supreme exponent of the subject is Euclid, whose authority remained almost unchallenged until very recent times.

It is probable that Euclid's own system of geometry is not now used in many schools, but children studying geometry become familiar with a number of the standard theorems and with the proofs of several of them. The general idea of a geometrical proof, if not of the details, will thus be familiar to anyone likely to read this book. We give in illustration a typical example, which we shall, in fact, find it necessary to criticise later. The proof will first be stated in standard form, and the nature of the geometrical arguments leading towards it will then be discussed.

To prove that the exterior angle of a triangle is greater than the interior opposite angles.

GIVEN: A triangle ABC whose side BC is produced beyond C to P (Fig. 6).

REQUIRED: To prove that the angle PCA is greater than the angle BAC.

CONSTRUCTION: Let O be the middle point of AC, and produce BO beyond O to D so that $OD = BO$.

PROOF: Since $OA = OC$ (construction)

$OB = OD$ (construction)

$\angle AOB = \angle COD$ (vertically opposite)

$\therefore\ \triangle AOB \equiv \triangle COD$ (SAS)

$\therefore\ \angle BAO = \angle DCO,$

that is, $\angle BAC = \angle DCA.$

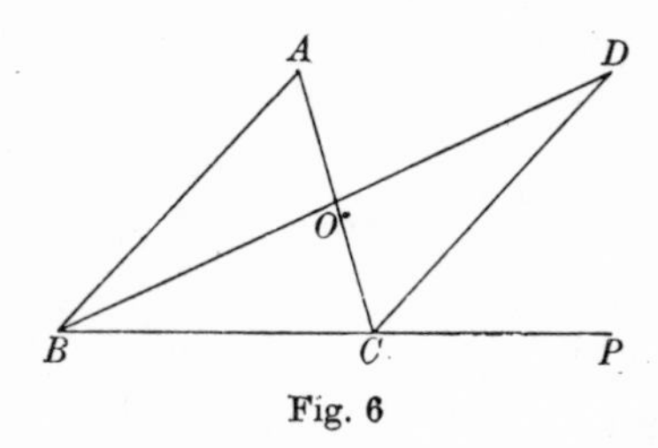

Fig. 6

But $\angle DCA$ is *part* of $\angle PCA$, so that

$$\angle DCA \text{ is less than } \angle PCA.$$

Hence also $\angle BAC$ is less than $\angle PCA$,

or $\angle PCA$ is greater than $\angle BAC$.

The essence of this proof (rarely used nowadays) is that it is enunciated as a necessary consequence of an earlier theorem, namely that two triangles are congruent which have two sides and the included angle equal. This theorem, in its turn, has been proved explicitly from yet earlier work (as in Euclid's treatment) or perhaps enunciated explicitly as one of the foundation stones on which the fabric of geometrical argument is to be built.

In this sort of way a structure is obtained involving point, line, angle, congruence, parallelism and so on, each new item being either defined explicitly as it appears or derived inductively from first principles or from preceding theory.

Since Euclid's time, his system of geometry has been subjected to very acute scrutiny, and a number of serious gaps have been revealed. In particular (and this is very relevant for the next chapter) it appears that he unwittingly made a number of assumptions from the appearance of his diagrams without realising that his logical foundations were thereby rendered suspect. The preceding illustration shows this very clearly.

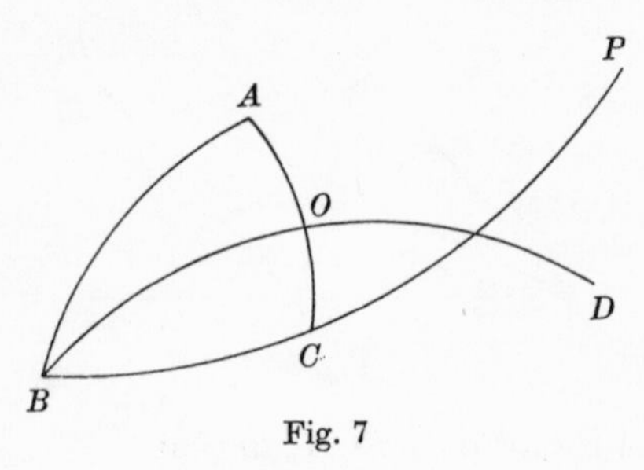

Fig. 7

The crux of the argument there is that the angle $\angle DCA$ is *part* of the angle $\angle PCA$, and this observation is usually followed in standard texts by the unassailable remark,

'but the whole is greater than its part',

so that the result follows. *There is, however, no step in the argument which proves from earlier results which is the whole and which the part.* The only reason for selecting $\angle PCA$ as the whole is that it 'looks like it'; but whether it would continue to do so for a triangle of atomic or astronomic dimensions is a very different matter. The result may, indeed, be untrue for a triangle drawn on a sphere (compare Fig. 7), so something is involved which is basically different for sphere and plane. The immediate point, however, is that this step of the argument is 'diagram', not logic.

Discussion of all that is involved would carry us far beyond the present aim. There is a very full treatment of the subject in *Foundations of Euclidean Geometry* by H. G. Forder (Cambridge University Press, 1927). We conclude this chapter by giving explicit mention to two basic geometrical properties not usually considered in elementary geometry, namely those covered by the words *between* and *outside* (or *inside*). These are the two characteristics of figures in space which remain undefined in the usual treatments, and the logical gaps left by their omission are precisely those which are often, unintentionally, taken over from a diagram. The work of the next chapter will reveal the damage that can be done by ignoring them.

CHAPTER IV

THE 'ISOSCELES TRIANGLE' FALLACY ANALYSED

It is customary to dispose of this fallacy by drawing an accurate diagram; but analysis by argument is fruitful, and we shall let the discussion lead us whither it will.

(i) Note first that the internal bisector of the angle A and the perpendicular bisector of BC both pass through the middle point of the arc BC opposite to A of the circumcircle of the triangle ABC (Fig. 8). In the earlier diagram (Fig. 2, p. 13), O was placed inside (in the normal intuitive sense of the word) the triangle; it is now seen to be outside, though, as we pointed out in ch. III, this fact cannot be established by any of the results of elementary geometry, which does not define the term.

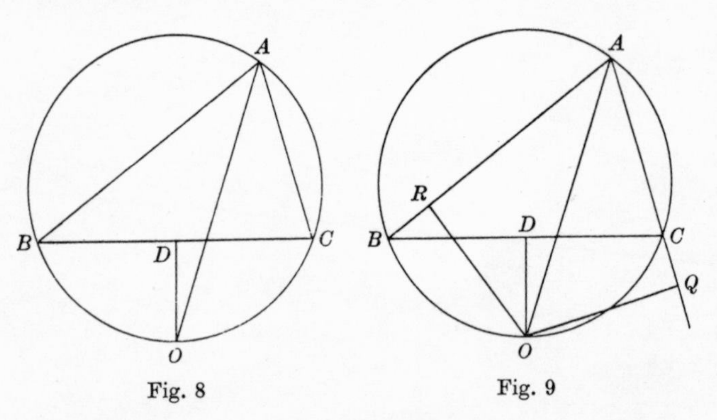

Fig. 8 Fig. 9

The removal of O to the outside of the triangle, however, does not of itself expose any fatal error, and we must proceed further. The points D, Q, R (Fig. 9) are now known to be the feet of the perpendiculars on the sides of a triangle *from a point O on the circumcircle.* Hence, by the Simson-line property,

$$D, Q, R \text{ are collinear.}$$

We therefore *appear* to have resolved the difficulty, since, for collinearity to be possible, one of the points Q, R must lie upon a side of the triangle and the other on a side produced. Thus (in the figure as drawn)

$$AB = AR + RB,$$

but
$$AC = AQ - QC.$$

The equality of AB and AC is thus negatived.

Once again, however, the axioms and theorems of elementary geometry are not sufficient to establish absolute proof. It is not possible to *prove* by them alone that a straight line cannot cut (internally) *all three* sides of a triangle. The final step requires a supplementary axiom, enunciated (like the other axioms) from experience and not from preceding logic:

PASCH'S AXIOM. *A straight line cutting one side of a triangle necessarily cuts one, and only one, of the other two sides, except in the case when it passes through the opposite vertex.*

The line through D therefore meets *either* AB *or* AC, but not both, and the discussion reaches its close.

The crux of this discussion is that the attempt to argue solely from the usual axioms and theorems has been found inadequate. It is necessary to supplement them either by reference to a diagram, or by Pasch's axiom. But reference to a diagram is unsatisfactory; it is hard enough in any case to be sure that possible alternatives have been excluded, and, even more seriously, there remains the criticism that something outside the logical structure has been called in to bolster it up. This, in a deliberately self-contained system like Euclidean geometry, cannot be allowed.

(ii) The introduction of trigonometry seems to give some sort of an answer to the problem, and is of interest.

Since
$$\angle OCA = C + \tfrac{1}{2}A,$$
application of the '$a/\sin A = 2R$' formula to the triangle OCA gives
$$\frac{OA}{\sin(C+\frac{1}{2}A)} = 2R,$$
where R is the radius of the circumcircle of the triangle ABC. Thus
$$\begin{aligned} AQ = AR = OA \cos\tfrac{1}{2}A \\ &= 2R\sin(C+\tfrac{1}{2}A)\cos\tfrac{1}{2}A \\ &= R\{\sin(C+A)+\sin C\} \\ &= R\{\sin B+\sin C\} \\ &= \tfrac{1}{2}\{2R\sin B+2R\sin C\} \\ &= \tfrac{1}{2}(b+c). \end{aligned}$$

Hence
$$\begin{aligned} AB-AR &= c-\tfrac{1}{2}(b+c) \\ &= \tfrac{1}{2}(c-b), \end{aligned}$$
and
$$\begin{aligned} AC-AQ &= b-\tfrac{1}{2}(b+c) \\ &= \tfrac{1}{2}(b-c), \end{aligned}$$
so that one of these quantities is positive and the other is negative; that is, one of the points Q, R lies on a side of the triangle and the other lies on a side produced—all without Pasch's axiom or any equivalent discussion.

The objection to this treatment lies concealed in the step
$$2\sin(C+\tfrac{1}{2}A)\cos\tfrac{1}{2}A = \sin(C+A)+\sin C,$$

since the proof of this formula involves ultimately the earlier formula '$\sin(A+B) = \sin A \cos B + \cos A \sin B$', and, through it, an appeal to a diagram to settle sense (positive or negative). A proof somewhat more exciting and illuminating than the standard one can be given by transferring the argument to Ptolemy's theorem, as in the next section.

(iii) Let PQ be a diameter of a circle of radius a and centre O (Fig. 10). Take points M, N on opposite sides of PQ so that (assuming the angles to be acute)

$$\angle QPM = A, \quad \angle QPN = B.$$

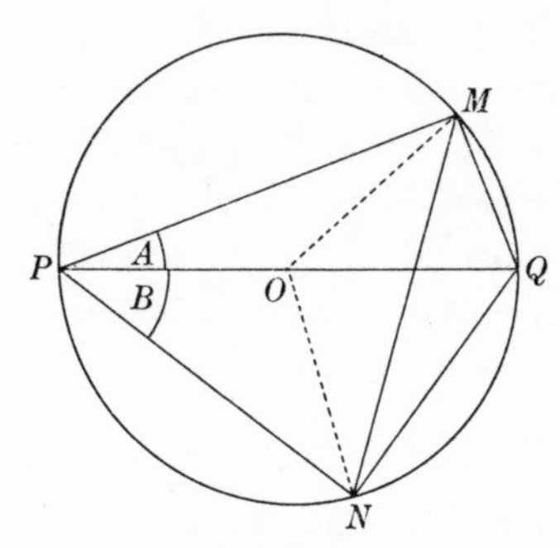

Fig. 10

Then $$\angle MON = 2(A+B).$$

Now $$PM = 2a\cos A, \quad QM = 2a\sin A,$$

$$PN = 2a\cos B, \quad QN = 2a\sin B,$$

$$MN = 2\{\tfrac{1}{2}MN\} = 2\{a\sin(A+B)\}$$

$$= 2a\sin(A+B).$$

But, by the theorem of Ptolemy,

$$PQ.MN = QM.PN + PM.QN,$$

so that

$$4a^2\sin(A+B) = 4a^2\sin A\cos B + 4a^2\cos A\sin B,$$

or $$\sin(A+B) = \sin A \cos B + \cos A \sin B.$$

This seems to establish the basic theorem authorising the step

$$2\sin(C+\tfrac{1}{2}A)\cos\tfrac{1}{2}A = \sin(C+A)+\sin C$$

in the preceding discussion.

We now give further attention to the theorem of Ptolemy. It can be restated in the form:

If A, B, C, D are four concyclic points and if the three products

$$BC.AD, \quad CA.BD, \quad AB.CD$$

are formed, then the sum of two of them is equal to the third.

The difficulty confronting us is to decide which is 'the third'. The usual method is to say that 'the third' is the product of the two *diagonals* of the cyclic quadrilateral whose vertices are A, B, C, D, but *this is precisely the appeal to diagram* that we seek to avoid. There are otherwise *three* distinct possibilities:

$$CA.BD+AB.CD = BC.AD,$$

$$AB.CD+BC.AD = CA.BD,$$

$$BC.AD+CA.BD = AB.CD,$$

and, for any given configuration, the criterion is the diagram. Without it we cannot tell which of the alternatives to select.

A proof of Ptolemy's theorem itself may be developed by a trigonometrical argument which leads directly to this dilemma. Let A, B, C, D be any four points on a circle of centre O and radius a (Fig. 11). Imagine a radius vector to rotate about O from the initial position OD in the counterclockwise sense, passing through A after an angle α, B after an angle β, and C after an angle γ, where α, β, γ are unequal, and all between 0 and 2π. It is not at present known in what order the points lie round the circle. Then, whether the angles α, β, γ are acute, obtuse or reflex, it is true that

$$AD = 2a \sin \tfrac{1}{2}\alpha,$$

$$BD = 2a \sin \tfrac{1}{2}\beta,$$

$$CD = 2a \sin \tfrac{1}{2}\gamma.$$

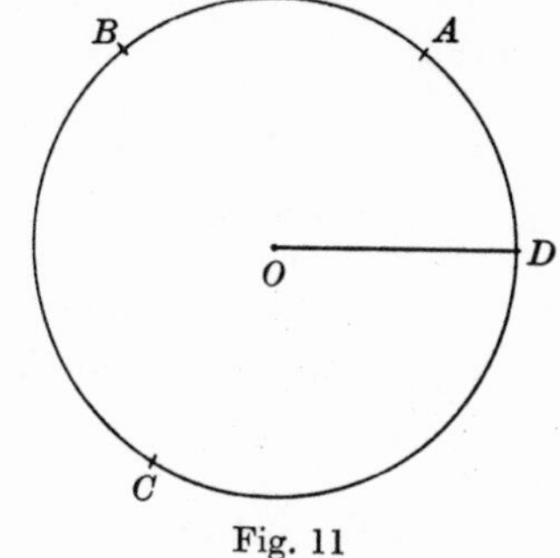

Fig. 11

Also

$$BC = 2a \sin \tfrac{1}{2}|\beta-\gamma|,$$

$$CA = 2a \sin \tfrac{1}{2}|\gamma-\alpha|,$$

$$AB = 2a \sin \tfrac{1}{2}|\alpha-\beta|,$$

where, for example, $|\beta-\gamma|$ means the *numerical value* of $(\beta-\gamma)$.

Now it is easy to establish by direct expansion the identity

$$\sin \tfrac{1}{2}\alpha \sin \tfrac{1}{2}(\beta-\gamma) + \sin \tfrac{1}{2}\beta \sin \tfrac{1}{2}(\gamma-\alpha) + \sin \tfrac{1}{2}\gamma \sin \tfrac{1}{2}(\alpha-\beta) = 0,$$

and the point to be made is that *this simple identity covers the three possible cases of the theorem of Ptolemy for four concylic points A, B, C, D*:

Of the three unequal angles α, β, γ, one must be intermediate in value between the other two; suppose that this is α. Then *either*

(i) $$\beta > \alpha > \gamma,$$

so that

$$|\beta-\gamma| = \beta-\gamma, \quad |\gamma-\alpha| = \alpha-\gamma, \quad |\alpha-\beta| = \beta-\alpha$$

and the identity is

$$\sin \tfrac{1}{2}\alpha \sin \tfrac{1}{2}|\beta-\gamma| - \sin \tfrac{1}{2}\beta \sin \tfrac{1}{2}|\gamma-\alpha| - \sin \tfrac{1}{2}\gamma \sin \tfrac{1}{2}|\alpha-\beta| = 0,$$

or $$AD.BC = BD.CA + CD.AB,$$

which is Ptolemy's theorem when AD, BC are the diagonals; *or*

(ii) $$\gamma > \alpha > \beta,$$

so that

$$|\beta-\gamma| = \gamma-\beta, \quad |\gamma-\alpha| = \gamma-\alpha, \quad |\alpha-\beta| = \alpha-\beta,$$

leading to the same relation

$$AD.BC = BD.CA + CD.AB.$$

When β is the intermediate angle the relation is

$$BD.CA = CD.AB + AD.BC$$

and when γ is the intermediate angle the relation is

$$CD.AB = AD.BC + BD.CA.$$

The three 'Ptolemy' possibilities are thus obtained, the concept of *diagonals* for the quadrangle being replaced by that of *between-ness* for the angles α, β, γ.

The point can be further emphasised by a more advanced argument from analytical geometry. Let the rectangular Cartesian coordinates of A, B, C, D be (x_1, y_1), (x_2, y_2), (x_3, y_3), (x_4, y_4), respectively, and denote by l_{ij} the distance between the points (x_i, y_i), (x_j, y_j). Consider the product of two determinants:

$$\begin{vmatrix} x_1^2+y_1^2 & -2x_1 & -2y_1 & 1 \\ x_2^2+y_2^2 & -2x_2 & -2y_2 & 1 \\ x_3^2+y_3^2 & -2x_3 & -2y_3 & 1 \\ x_4^2+y_4^2 & -2x_4 & -2y_4 & 1 \end{vmatrix}$$

$$\times \begin{vmatrix} 1 & 1 & 1 & 1 \\ x_1 & x_2 & x_3 & x_4 \\ y_1 & y_2 & y_3 & y_4 \\ x_1^2+y_1^2 & x_2^2+y_2^2 & x_3^2+y_3^2 & x_4^2+y_4^2 \end{vmatrix}.$$

On multiplication according to the 'matrix' rule the element in the ith row and jth column is

$$(x_i^2+y_i^2)-2x_ix_j-2y_iy_j+(x_j^2+y_j^2),$$

or

$$(x_i-x_j)^2+(y_i-y_j)^2,$$

or

$$l_{ij}^2,$$

Hence the product is

$$\begin{vmatrix} 0 & l_{12}^2 & l_{13}^2 & l_{14}^2 \\ l_{21}^2 & 0 & l_{23}^2 & l_{24}^2 \\ l_{31}^2 & l_{32}^2 & 0 & l_{34}^2 \\ l_{41}^2 & l_{42}^2 & l_{43}^2 & 0 \end{vmatrix}.$$

This determinant may be evaluated by direct computation. Alternatively, write

$$l_{23}^2 = l_{32}^2 = f;\quad l_{31}^2 = l_{13}^2 = g;\quad l_{12}^2 = l_{21}^2 = h;$$
$$l_{14}^2 = l_{41}^2 = l;\quad l_{24}^2 = l_{42}^2 = m;\quad l_{34}^2 = l_{43}^2 = n,$$

and replace zeros temporarily by the letters a, b, c. Then the determinant is

$$\begin{vmatrix} a & h & g & l \\ h & b & f & m \\ g & f & c & n \\ l & m & n & 0 \end{vmatrix},$$

which, equated to zero, may be recognised as giving the *tangential equation* of the conic

$$ax^2+by^2+cz^2+2fyz+2gzx+2hxy = 0,$$

namely

$$Al^2+Bm^2+Cn^2+2Fmn+2Gnl+2Hlm = 0,$$

where $\qquad A = bc-f^2, \quad F = gh-af,$

or (with a, b, c returned to zero)

$$-f^2l^2-g^2m^2-h^2n^2+2ghmn+2hfnl+2fglm = 0.$$

Further, the value of the determinant is indeed zero if A, B, C, D are taken to be concyclic, the condition for this being

$$\begin{vmatrix} x_1^2+y_1^2 & x_1 & y_1 & 1 \\ x_2^2+y_2^2 & x_2 & y_2 & 1 \\ x_3^2+y_3^2 & x_3 & y_3 & 1 \\ x_4^2+y_4^2 & x_4 & y_4 & 1 \end{vmatrix} = 0.$$

(This is found by substituting the coordinates of A, B, C, D in the standard equation

$$x^2+y^2+2gx+2fy+c = 0$$

for a circle, and then eliminating determinantally the ratios

$$1:2g:2f:c.)$$

The condition for the four points to be concyclic is thus

$$f^2l^2+g^2m^2+h^2n^2-2ghmn-2hfnl-2fglm = 0,$$

or

$$\pm\sqrt{(fl)}\pm\sqrt{(gm)}\pm\sqrt{(hn)} = 0.$$

(This is a well-known formula for the tangential equation of a conic through the vertices of the triangle of reference.)

In terms of l_{ij}, the condition is

$$\pm\sqrt{(l_{23}^2 l_{14}^2)}\pm\sqrt{(l_{31}^2 l_{24}^2)}\pm\sqrt{(l_{12}^2 l_{34}^2)} = 0,$$

or

$$\pm l_{23}l_{14}\pm l_{31}l_{24}\pm l_{12}l_{34} = 0.$$

This gives the three distinct possibilities

$$l_{31}l_{24}+l_{12}l_{34} = l_{23}l_{14},$$
$$l_{12}l_{34}+l_{23}l_{14} = l_{31}l_{24},$$
$$l_{23}l_{14}+l_{31}l_{24} = l_{12}l_{34},$$

remembering that the distances are positive so that the three alternative signs cannot be all alike.

We have therefore recovered the three relations given on p. 28, and this is all that can be done without further information about relative positions round the circle.

CHAPTER V

THE OTHER GEOMETRICAL FALLACIES ANALYSED

THE FALLACY OF THE RIGHT ANGLE. Here again a proof of the point at which fallacy intrudes leads to considerations of some depth. We seek as a preliminary step an alternative interpretation for the position of the point O (p. 14).

We have seen that PO, RO are the perpendicular bisectors of CD, DE, and so *O is the circumcentre of the triangle CDE*. Hence O lies also on the perpendicular bisector of CE.

But, by construction, $BC = BE$, so that the perpendicular bisector of CE also passes through B. Moreover, and this cannot be proved by the usual axioms and theorems alone, this line is the *internal* bisector of the angle $\angle CBE$. Thus OB passes down the internal bisector of the angle $\angle CBE$, so that *the angle $\angle OBE$ is reflex*, being equal to the sum of two right angles plus $\frac{1}{2}\angle CBE$. The triangle OBE is therefore so placed that OE is on the side of B remote from the rest of the figure. The step

$$\angle ABE = \angle OBE \pm \angle OBA$$

is therefore illegitimate.

Note that the exposure of the fallacy has involved arguments not supported by the usual axioms and theorems. It has been necessary to consider the internal bisector of an angle; and ideas of *inside* and *outside* appear in Euclid by reference to diagrams and not by reason only, as we have already suggested in ch. III.

THE TRAPEZIUM FALLACY. The argument here is unexpectedly close to that of the preceding Fallacy. It is, of course, perfectly possible for the result to be true; we are interested only in the cases where it is not. In such a case, draw the line through B parallel to AD (Fig. 12). Let the circle with centre D and radius $AB = CD$ meet this line in F and in G (where G, not shown in the diagram, is the point such that $ABGD$ is a parallelogram), with DF not parallel to AB.

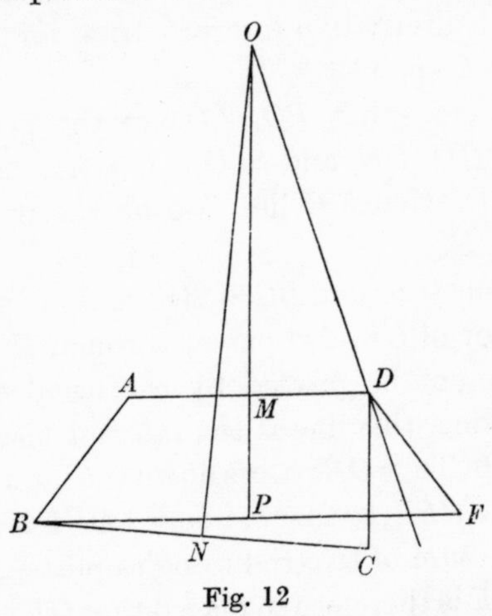

Fig. 12

Then by symmetry (or by easy argument) OM is also the perpendicular bisector of BF; and ON is the perpendicular bisector of BC. Hence *O is the circumcentre of the triangle BCF*, so that, as in the preceding Fallacy, OD lies along the internal bisector of the angle CDF, and the exposure follows as before.

THE FALLACY OF THE EMPTY CIRCLE. The fallacy is most convincingly exposed by algebra. Let

$$OP = p,$$

so that

$$OQ = r^2/p.$$

Now $$(r-p)^2 > 0,$$
since the left-hand side is a perfect square. Hence
$$r^2-2rp+p^2 > 0,$$
or $$p^2+r^2 > 2rp,$$
or (dividing by the positive number $2p$)
$$\tfrac{1}{2}(p+r^2/p) > r.$$
Hence $$OR > r,$$
that is, the point R lies *outside* the circle, and so the points U, V do not exist and no calculations can be made involving them.

The result is, however, startling when approached with the technique of coordinate geometry. Take OP to be the x-axis and the line through O perpendicular to OP to be the y-axis. Then P is the point $(p, 0)$, Q is the point $(r^2/p, 0)$ and R is the point $\{\tfrac{1}{2}(r^2/p+p), 0\}$.

The circle is $$x^2+y^2 = r^2,$$
and this cuts the line through R perpendicular to OP, that is, the line $$x = \tfrac{1}{2}(r^2/p+p),$$
where $$y^2 = r^2-\tfrac{1}{4}(r^2/p+p)^2$$
$$= -\tfrac{1}{4}(r^2/p-p)^2.$$

In a real geometry no such value of y exists and the argument stops. We therefore continue it in *complex Cartesian geometry*. Then
$$y = \pm\tfrac{1}{2}i(r^2/p-p),$$
where $i = \sqrt{(-1)}$. Thus the complex perpendicular bisector of PQ cuts the complex circle in two points, of which a typical one is
$$U\{\tfrac{1}{2}(r^2/p+p), \tfrac{1}{2}i(r^2/p-p)\}.$$

Now apply the formula
$$d^2 = (x_1-x_2)^2+(y_1-y_2)^2$$

to calculate the distance PU. Thus

$$PU^2 = \{\tfrac{1}{2}(r^2/p+p)-p\}^2+\{\tfrac{1}{2}i(r^2/p-p)-0\}^2$$
$$= \tfrac{1}{4}(r^2/p-p)^2+\tfrac{1}{4}i^2(r^2/p-p)^2$$
$$= \tfrac{1}{4}(r^2/p-p)^2(1+i^2)$$
$$= 0,$$

so that *the distance between P and U is zero.*

(This is less odd, however, than it sounds, for it depends to a large extent on the transference of the word 'distance' to this context.)

The circle of centre P and passing through U is thus given by the equation

$$(x-p)^2+y^2 = 0,$$

and this meets the given circle

$$x^2+y^2 = r^2$$

on their radical axis

$$\{(x-p)^2+y^2\}-\{x^2+y^2-r^2\} = 0,$$

or $$-2px+p^2+r^2 = 0,$$

or $$x = \tfrac{1}{2}(r^2/p+p),$$

which is precisely the line UV already considered.

Thus there are two points on the circle whose distance from P in complex Cartesian geometry is zero. They are, however, quite distinct from P.

Corollary: given any circle and any point in its plane, there exist in complex Cartesian geometry two points on the circle whose distance from that point is zero, where distance is defined by the formula

$$d^2 = (x_1-x_2)^2+(y_1-y_2)^2.$$

Note that this corollary, as stated, is not a fallacy, but is inherent in the definition of distance for complex geometry.

CHAPTER VI

SOME FALLACIES IN ALGEBRA AND TRIGONOMETRY

We gather together here some fallacies based on a disregard of the more elementary rules of algebraic and trigonometric manipulation. As the reader may prefer not to have the exposure immediately beside the fallacy itself, we reserve the comment (here and in later fallacies) till the end of the chapter.

I. The Fallacies

¶ 1. The Fallacy that $4 = 0$.

Since

$$\cos^2 x = 1 - \sin^2 x,$$

it follows that

$$1 + \cos x = 1 + (1 - \sin^2 x)^{\frac{1}{2}},$$

or, squaring each side,

$$(1 + \cos x)^2 = \{1 + (1 - \sin^2 x)^{\frac{1}{2}}\}^2.$$

In particular, when $x = \pi$,

$$(1 - 1)^2 = \{1 + (1 - 0)^{\frac{1}{2}}\}^2,$$

or

$$0 = (1 + 1)^2$$

$$= 4.$$

¶ 2. The Fallacy that $+1 = -1$.

Since

$$1 = \sqrt{1}$$

$$= \sqrt{\{(-1)(-1)\}}$$

$$= \sqrt{(-1)}\sqrt{(-1)},$$

it follows, writing $\sqrt{(-1)} \equiv i$, that

$$\begin{aligned} 1 &= i.i \\ &= i^2 \\ &= -1. \end{aligned}$$

The two fallacies which follow are somewhat similar at the critical point. It is perhaps worth while to include both for the sake of the geometry with which they are clothed.

¶ 3. The Fallacy that All Lengths are Equal.

GIVEN: A, C, D, B are four points in order on a straight line.

REQUIRED: To prove that $AC = BD$, necessarily.

CONSTRUCTION (Fig. 13): Let O be one of the points of intersection of the perpendicular bisector of CD with the circle of centre A and radius $\sqrt{(AC.AB)}$, so that

$$AO^2 = AC.AB.$$

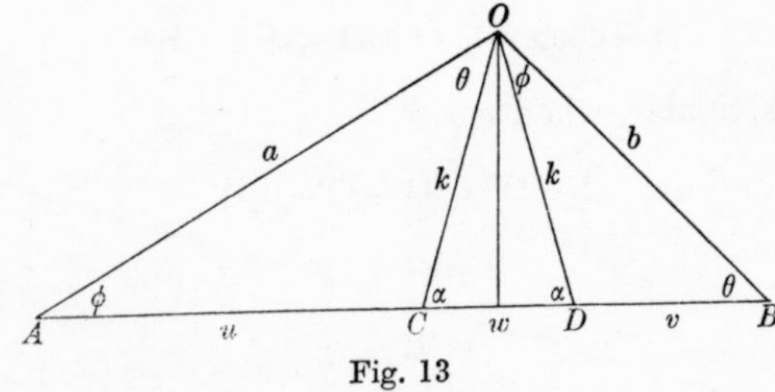

Fig. 13

PROOF: Since $AO^2 = AC.AB$, the line AO is the tangent at O to the circle OCB,

$$\begin{aligned} \therefore \angle AOC &= \angle CBO \\ &= \theta, \quad \text{say.} \end{aligned}$$

Since $$OC = OD$$

$$\angle OCD = \angle ODC$$

$$= \alpha, \quad \text{say},$$

$$\therefore \angle OAC = \alpha - \theta = \angle BOD$$

$$= \phi, \quad \text{say}.$$

Write $OA = a$, $OB = b$; $AC = u$, $BD = v$, $CD = w$; $OC = OD = k$.

From $\triangle OAC$,

$$a^2 = k^2 + u^2 + 2ku \cos \alpha.$$

From $\triangle OBD$,

$$b^2 = k^2 + v^2 + 2kv \cos \alpha.$$

Hence $$(a^2 - k^2 - u^2)\, v = 2kuv \cos \alpha$$

$$= (b^2 - k^2 - v^2)\, u,$$

so that $$(k^2 + v^2)\, u - (k^2 + u^2)\, v = b^2 u - a^2 v,$$

or $$(u - v)(k^2 - uv) = b^2 u - a^2 v,$$

or $$u - v = \frac{b^2 u - a^2 v}{k^2 - uv}.$$

Now the triangles OAC, BOD are similar, so that

$$\frac{a^2}{b^2} = \frac{\triangle OAC}{\triangle BOD} = \frac{AC}{BD} = \frac{u}{v}.$$

Hence $$b^2 u - a^2 v = 0$$

$$\therefore u - v = 0,$$

or $$u = v.$$

¶ 4. The Fallacy that Every Triangle is Isosceles.

GIVEN: A triangle ABC.

REQUIRED: To prove that $AB = AC$, necessarily.

CONSTRUCTION: Draw the bisector of the angle A to meet BC at D, and produce AD to P so that

$$AD.DP = BD.DC$$

(Fig. 14).

PROOF: For convenience of notation, write $AB = c$, $AC = b$, $BD = u$, $CD = v$, $AD = x$, $PD = y$, $PB = r$, $PC = q$, $\angle PDB = \theta$. The definition of P gives the relation

$$xy = uv.$$

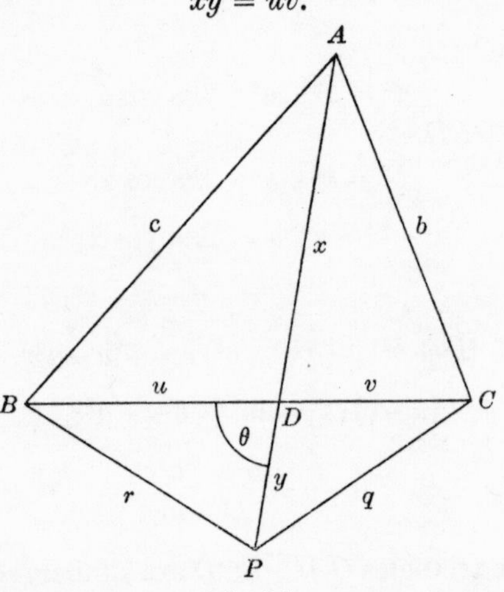

Fig. 14

Consider the triangles ADC, BDP:

$$\angle ADC = \angle BDP$$

$$\frac{AD}{DC} = \frac{BD}{DP}. \qquad \text{(contruction)}$$

$\therefore$ The triangles are similar (2 sides about equal angles proportional).

$$\therefore \frac{AD}{BD} = \frac{DC}{DP} = \frac{AC}{BP},$$

or

$$\frac{x}{u} = \frac{v}{y} = \frac{b}{r}.$$

Similarly the triangles ADB, CDP are similar, and

$$\frac{x}{v} = \frac{u}{y} = \frac{c}{q}.$$

In particular,
$$\frac{v}{y}.\frac{u}{y} = \frac{b}{r}.\frac{c}{q},$$

or
$$uvqr = bcy^2.$$

Now the length of the bisector AD is given by the standard theorem

$$AB.AC = AD^2 + BD.DC$$

$$\therefore\ bc = x^2 + uv.$$

Hence
$$\begin{aligned} uvqr &= x^2y^2 + uvy^2 \\ &= u^2v^2 + uvy^2 \end{aligned}$$

$$\therefore\ qr = uv + y^2.$$

Also, from the triangle PBD,

$$r^2 = u^2 + y^2 - 2uy\cos\theta,$$

and, from the triangle PCD,

$$q^2 = v^2 + y^2 + 2vy\cos\theta.$$

Multiply these equations by v, u, respectively, and add:

$$\begin{aligned} vr^2 + uq^2 &= uv(u+v) + y^2(u+v) \\ &= (u+v)(uv+y^2) \\ &= (u+v)\,qr, \end{aligned}$$

by the previous result.

Let us now rearrange this relation to give the ratio u/v:

$$u(q^2 - qr) = v(qr - r^2),$$

or
$$\begin{aligned} \frac{u}{v} &= \frac{qr - r^2}{q^2 - qr} = \frac{r(q-r)}{q(q-r)} \\ &= \frac{r}{q}. \end{aligned}$$

That is, $$\frac{DB}{DC} = \frac{PB}{PC},$$

so that *PD bisects the angle BPC*.

Finally, compare the triangles APB, APC:

$$AP = AP,$$

$$\angle BAP = \angle CAP \qquad \text{(construction)}$$

$$\angle APB = \angle APC \qquad \text{(proved)}$$

$$\therefore \triangle APB \equiv \triangle APC. \qquad \text{(ASA)}$$

In particular, $$AB = AC.$$

5. The Fallacy that $+1 = -1$.

To solve the equation $\cot\theta + \tan 3\theta = 0$.

$$\cot\theta + \tan(\theta + 2\theta) = 0$$

$$\therefore \cot\theta + \frac{\tan\theta + \tan 2\theta}{1 - \tan\theta\tan 2\theta} = 0$$

$$\therefore \cot\theta - \tan 2\theta + \tan\theta + \tan 2\theta = 0$$

$$\therefore \cot\theta + \tan\theta = 0$$

$$\therefore \tan^2\theta + 1 = 0$$

$$\therefore \tan^2\theta = -1$$

$$\therefore \tan\theta = \pm i \qquad (i = \sqrt{(-1)}).$$

But there are genuine values* for θ, in the form

$$\theta = \tfrac{1}{4}\pi + \tfrac{1}{2}n\pi \qquad (n \text{ integer}).$$

---

* $$\tan 3\theta = -\cot\theta$$
$$= \tan(\theta + \tfrac{1}{2}\pi)$$
$$\therefore 3\theta = \theta + \tfrac{1}{2}\pi + n\pi$$
$$\therefore \theta = \tfrac{1}{4}\pi + \tfrac{1}{2}n\pi.$$

For example, $$\tan \tfrac{3}{4}\pi = -\cot\tfrac{1}{4}\pi.$$

Thus, with, say, $$\theta = \tfrac{1}{4}\pi,$$

$$\tan^2 \tfrac{1}{4}\pi = -1$$

$$\therefore \quad +1 = -1.$$

❡ 6. THE FALLACY THAT EVERY ANGLE IS A MULTIPLE OF TWO RIGHT ANGLES.

Let θ be an angle (complex) satisfying the relation

$$\tan\theta = i.$$

Then, if A is any angle,

$$\tan(A+\theta) = \frac{\tan A + \tan\theta}{1 - \tan A \tan\theta}$$

$$= \frac{\tan A + i}{1 - i \tan A}$$

$$= i$$

$$= \tan\theta.$$

Thus $$\tan(A+\theta) = \tan\theta,$$

so that $$A + \theta = n\pi + \theta,$$

or $$A = n\pi$$

for any angle A.

❡ 7. THE FALLACY THAT $\pi = 0$.

Since $$e^{2\pi i} = \cos 2\pi + i \sin 2\pi$$

$$= 1,$$

it follows that, for any value of x,

$$e^{ix} = e^{ix}\, e^{2\pi i} = e^{i(x+2\pi)}.$$

Raise each side to the power i:

$$(e^{ix})^i = (e^{i(x+2\pi)})^i,$$

or $$e^{-x} = e^{-(x+2\pi)}.$$

Multiply each side by $e^{x+2\pi}$, which cannot be zero for any value of x. Then

$$e^{2\pi} = 1,$$

so that

$$2\pi = 0.$$

¶ 8. THE FALLACY THAT THE SUM OF THE SQUARES ON TWO SIDES OF A TRIANGLE IS NEVER LESS THAN THE SQUARE ON THE THIRD.

GIVEN: A triangle ABC so named that

$$a > b.$$

REQUIRED: To prove that $a^2+b^2 > c^2$, necessarily.

PROOF: Since

$$a > b,$$

it follows that

$$a\cos C > b\cos C.$$

But, by standard formula,

$$a = b\cos C + c\cos B \quad \therefore\ b\cos C = a - c\cos B,$$

$$b = a\cos C + c\cos A \quad \therefore\ a\cos C = b - c\cos A.$$

Hence

$$b - c\cos A > a - c\cos B$$

$$\therefore\ c\cos B - c\cos A > a - b.$$

Multiply each side by $2ab$ and use the cosine formula.

$$\therefore\ b(a^2+c^2-b^2) - a(b^2+c^2-a^2) > 2ab(a-b)$$

$$\therefore\ a^3-b^3-a^2b+ab^2 > c^2(a-b),$$

$$\therefore\ (a-b)(a^2+b^2) > c^2(a-b),$$

$$\therefore\ a^2+b^2 > c^2,$$

division by $a-b$ being legitimate since $a-b>0$.

ILLUSTRATION. Consider the particular case

$$a = 4, \quad b = 3, \quad c = 6.$$

This triangle exists, since

$$b+c > a, \quad c+a > b, \quad a+b > c.$$

But $$a^2+b^2 = 25, \quad c^2 = 36.$$

Hence, by the theorem,

$$25 > 36.$$

II. The Commentary

Fallacies 1 and 2.

These are both based on the ambiguity of sign which arises whenever square roots are taken. An equation

$$x^2 = a^2$$

has two solutions,

$$x = +a, \quad x = -a,$$

and care must always be taken to select the appropriate one. It is not necessarily true that both solutions are relevant in the problem giving rise to the equation, and it is *always* essential to check independently.

This phenomenon is well known in the case of the solution of equations involving surds. Such equations can, indeed, be stated to give an elementary form of fallacy:

To prove that $1 = 3$.

Consider the equation

$$\sqrt{(5-x)} = 1+\sqrt{x}.$$

Square each side:

$$5-x = 1+2\sqrt{x}+x,$$

or $$4-2x = 2\sqrt{x},$$

or $$2-x = \sqrt{x}.$$

Square each side: $$4-4x+x^2 = x,$$

or $$x^2-5x+4 = 0,$$

or $$(x-4)(x-1) = 0,$$

so that $$x = 4 \text{ or } 1.$$

In particular, the value $x = 4$ gives, on substitution in the given equation,
$$\sqrt{(5-4)} = 1+\sqrt{4},$$

or $$1 = 1+2$$
$$= 3.$$

The fact is that the whole of the argument has moved in one direction only, and there are several points where it cannot be reversed. The step
$$5-x = 1+2\sqrt{x}+x^2$$
is a consequence not only of the given equation but also of the distinct equation
$$-\sqrt{(5-x)} = 1+\sqrt{x}.$$
We have in effect solved *four* equations,
$$\pm\sqrt{(5-x)} = 1\pm\sqrt{x},$$
and each of the two prospective solutions must be checked against the equation actually proposed.

The following synthetic example of false square roots affords another exceedingly simple illustration:

By direct computation of each side,
$$9-24 = 25-40.$$
Add 16 to each side:
$$9-24+16 = 25-40+16$$
$$\therefore\ (3-4)^2 = (5-4)^2.$$
Take square roots: $$\therefore\ 3-4 = 5-4.$$
Hence $$3 = 5.$$

With these remarks in mind, the reader should have no difficulty in disposing of Fallacy 1. The second one is, however, worthy of further comment.

The steps
$$1 = \sqrt{1}$$
$$= \sqrt{\{(-1)(-1)\}}$$

are correct. The following step

$$\sqrt{\{(-1)(-1)\}} = \sqrt{(-1)}\sqrt{(-1)}$$

may be regarded as correct, but it needs care in interpretation. The next step

$$\sqrt{(-1)}\sqrt{(-1)} = i.i$$

is definitely wrong, for each of the square roots $\sqrt{(-1)}$ and $\sqrt{(-1)}$ has, in the first instance, two possible values, and *the ambiguities cannot be resolved without reference to the rest of the problem.* If the symbol i is used for, say, the first square root, then it is in the nature of the problem that the second square root is necessarily $-i$.

This may be made clearer by examination of the exponential forms of the square roots. It is well known that, if n is any integer,

$$\begin{aligned} e^{(\frac{1}{2}+2n)\pi i} &= \cos(\tfrac{1}{2}+2n)\pi + i\sin(\tfrac{1}{2}+2n)\pi \\ &= \cos\tfrac{1}{2}\pi + i\sin\tfrac{1}{2}\pi \\ &= i, \end{aligned}$$

and that, if m is any integer, then, similarly,

$$e^{(-\frac{1}{2}+2m)\pi i} = -i.$$

If, in particular, we use the identification (with $n = 0$)

$$i = e^{\frac{1}{2}\pi i},$$

then the argument is

$$\begin{aligned} 1 &= \sqrt{(-1)}\sqrt{(-1)} \\ &= e^{\frac{1}{2}\pi i} \times (?), \end{aligned}$$

and it is at once evident that the disputed square root is $e^{-\frac{1}{2}\pi i}$, or $-i$, in agreement with the formula

$$\begin{aligned} 1 &= e^0 \\ &= e^{\frac{1}{2}\pi i}\, e^{-\frac{1}{2}\pi i}. \end{aligned}$$

FALLACIES 3 AND 4.

The resolution of these fallacies depends on the illegitimate step usually referred to as *division by zero*. It is usually true that, if

$$am = an,$$

then

$$m = n;$$

but this need not be so if a is zero. For example,

$$0.5 = 0.3,$$

but

$$5 \neq 3.$$

The division of each side of the equation by a is proper if, and only if, a is not zero.

There is an alternative way of regarding this process which is perhaps more significant. We may call it the *method of the false factor*. The relation

$$am = an$$

may be expressed in factorised form

$$a(m-n) = 0,$$

from which it follows that one at least of a, $m-n$ is zero. If, therefore, a is not zero, then $m-n$ must be.

The point is that the reader is presented with two factors and then subjected to pressure to accept the wrong one.

There is a crude type of fallacy based on division by zero, of which the following example is typical:

Let

$$x = 2.$$

Multiply each side by $x-1$:

$$x^2 - x = 2x - 2.$$

Subtract x from each side:

$$x^2 - 2x = x - 2.$$

Divide each side by $x-2$:

$$x = 1.$$

But it is given that $x = 2$. Hence

$$2 = 1.$$

The forced insertion of the factor $x-1$ is, however, very artificial. In the given fallacies the alternative factor arises in its own right out of the geometry and must be accounted for.

Note that the extraction of square roots referred to above is itself a particular case of false factors; for the equation

$$x^2 = a^2$$

may be written $$(x-a)(x+a) = 0,$$

and here again the wrong factor is insinuated into the argument.

Having these ideas in mind we turn to Fallacy 3. We reached correctly the step

$$(u-v)(k^2-uv) = b^2u-a^2v,$$

and later the equally correct step

$$b^2u-a^2v = 0.$$

Hence it is true that

$$(u-v)(k^2-uv) = 0.$$

The reader was then presented with the relation

$$u-v = 0,$$

but the alternative

$$k^2-uv = 0$$

also required consideration. Now the triangles OAC, BOD are similar, so that

$$\frac{OC}{BD} = \frac{AC}{OD},$$

or $$\frac{k}{v} = \frac{u}{k}.$$

The relation $$k^2 - uv = 0$$

is therefore true always, so that the deduction

$$u - v = 0$$

cannot be made.

The reader interested in geometry will notice that the relation

$$k^2 = uv$$

is an extension of a well-known result of elementary geometry. If $\angle AOB$ is a right angle, then C and D coincide at the foot of the altitude from O and

$$OC^2 = OD^2 = CA.CB.$$

The figure under consideration is, so to speak, a 'widening out' of the more familiar one.

The algebra in Fallacy 4 is surprisingly similar to that of number 3, but the underlying geometry has an interest of its own. The step

$$u(q^2 - qr) = v(qr - r^2)$$

is correct, and leads to the factorised form

$$(uq - vr)(q - r) = 0,$$

from which attention was drawn to the factor $uq - vr$. The possibility $q - r = 0$ must, however, be considered too. Now the relation

$$xy = uv$$

leads to the result that the quadrilateral $ABPC$ is cyclic, so that, since $\angle BAP = \angle CAP$, we have

$$BP = CP.$$

Hence, necessarily,

$$q - r = 0,$$

and the deduction $uq - vr = 0$ is thus inadmissible.

We may now examine the basic geometry behind this algebraic façade. The primary and standard theorem is that, *if the internal bisector of the angle BAC of a triangle meets the base in D, then*

$$AB.AC = AD^2 + BD.DC.$$

The problem (which we applied to the triangle PBC) is to determine whether the *converse* is also true, that the existence of a point D for which the above relation holds necessarily implies that AD bisects the angle A. The answer is that in general it does, but the argument breaks down if the triangle ABC is isosceles with $AB = AC$. In the latter case the relation holds for *all* positions of D on BC. By ignoring the fact that the triangle PBC is isosceles, we obtained PD as the angle bisector, leading at once to the fallacy.

FALLACIES 5 AND 6.

These two fallacies are similar in essential point, but it is interesting to remark that the first of them arose as an attempt to solve the given equation by a perfectly reasonable method. One suspects that the trap is one that might deceive many experienced mathematicians.

Suppose that any number k, real or complex, is given. Then, with proper definition of complex circular functions, it is always possible to solve the equation

$$\tan x = k,$$

since then

$$\frac{\sin x}{k} = \frac{\cos x}{1} = \frac{1}{\pm\sqrt{(1+k^2)}}.$$

There is, however, one case of exception, namely that which arises when

$$k^2 + 1 = 0,$$

or

$$k = \pm i.$$

Thus no (complex) angle exists whose tangent has either of the values $\pm i$.

Fallacy 6 therefore collapses completely, since the angle θ is non-existent, and Fallacy 5 must have an omission at some stage, since the step

$$\tan^2\theta + 1 = 0$$

is excluded.

The omission is one that might easily pass detection were it not for the fallacy to which it leads. There is in fact a solution $\theta = \frac{1}{4}\pi$, so that $2\theta = \frac{1}{2}\pi$; and $\tan 2\theta$ has no value (or, as is sometimes said, is infinite). The phenomenon may be exhibited more easily by taking the equation in its 'inverse' form

$$\cot\theta = -\tan 3\theta$$

$$\therefore\ \tan\theta = -\cot 3\theta$$

$$\therefore\ \tan\theta + \cot 3\theta = 0.$$

This gives

$$\tan\theta + \frac{\cot\theta\cot 2\theta - 1}{\cot\theta + \cot 2\theta} = 0$$

$$\therefore\ 1 + \tan\theta\cot 2\theta + \cot\theta\cot 2\theta - 1 = 0,$$

so that $$\cot 2\theta(\tan\theta + \cot\theta) = 0.$$

Since $\tan\theta + \cot\theta \neq 0$, the solution is

$$\cot 2\theta = 0,$$

or $$2\theta = \tfrac{1}{2}\pi + n\pi,$$

or $$\theta = \tfrac{1}{4}\pi + \tfrac{1}{2}n\pi.$$

The zero value of $\cot 2\theta$ corresponds to the 'infinite' value of $\tan 2\theta$.

FALLACY 7.

This fallacy reinforces the lesson of Fallacies 1, 2, that unless n is an integer (positive or negative), then the power

$$x^n$$

is a many-valued function; the particular value must therefore be defined *explicitly* when this becomes necessary. Thus the step

$$e^{ix} = e^{i(x+2\pi)}$$

is correct, but examination is required when raising either side to power i. Since

$$e^{ix} = e^{i(x+2p\pi)} \quad (p \text{ integral}),$$

the function $$(e^{ix})^i$$

has the infinite set of values

$$e^{-x}.e^{-2p\pi} \quad (p = \ldots, -2, -1, 0, 1, 2, \ldots).$$

The function $$(e^{i(x+2\pi)})^i$$

has similarly the infinite set of values

$$e^{-(x+2\pi)}.e^{-2q\pi} \quad (q = \ldots, -2, -1, 0, 1, 2, \ldots).$$

When the values of p, q are chosen correctly the paradox disappears. The most obvious choice is $p = 0$, $q = -1$, giving, impeccably,

$$e^{-x} = e^{-(x+2\pi)}.e^{2\pi}.$$

FALLACY 8.

Everybody who devotes any time at all to the subject knows that he must exercise the greatest care when multiplying the two sides of an inequality by some given number. For example, it is true that

$$5 > 4,$$

but not necessarily true that

$$5a > 4a.$$

If $a = 3$, the relation is

$$15 > 12,$$

which is correct; if $a = -3$, the relation is

$$-15 > -12,$$

which is not correct. It is necessary to make sure that the multiplier is *positive*.

Most inequality fallacies are based, rather tediously, on this principle, but one example ought to be included in the collection. The present illustration has at any rate the virtue of emphasising, what is well known otherwise, that $a^2 + b^2$ is greater than c^2 when the angle C is acute and less than c^2 when C is obtuse.

CHAPTER VII

FALLACIES IN DIFFERENTIATION

The following fallacies illustrate one or two points of danger in the use of the differential calculus. As before, comments are given in the later part of this chapter.

I. THE FALLACIES

¶ 1. THE FALLACY THAT THERE IS NO POINT ON THE CIRCUMFERENCE OF A CIRCLE NEAREST TO A POINT INSIDE IT.

Choose the given inside point as the origin O of Cartesian coordinates and the x-axis as the line joining O to the centre C. Let the coordinates of C then be $(a, 0)$ and the radius b, so that the equation of the circle is

$$(x-a)^2+y^2 = b^2,$$

or

$$x^2+y^2-2ax+a^2-b^2 = 0.$$

The distance from O of the point $P(x, y)$ of the circle is r, where

$$r^2 = x^2+y^2,$$

so that, since P is on the circle,

$$r^2 = 2ax-a^2+b^2.$$

The distance OP is least, or greatest, at points for which $\frac{dr}{dx} = 0$. But, differentiating the relation for r^2, we have

$$r\frac{dr}{dx} = a.$$

Now r is not zero or 'infinity', and this equation can only be satisfied when $\frac{dr}{dx} = 0$ if $a = 0$, that is, if O is at the centre of the circle—in which case all points of the circumference, being at constant distance from O, are alike nearest to it and furthest from it.

If, as is true in general, a is not zero, then there is no point on the circumference whose distance from O is either a maximum or a minimum.

¶ 2. The Fallacy of the Radius of a Circle.

To prove that the radius of a circle is indeterminate.

The working is based on the standard formula

$$\rho = r\frac{dr}{dp}$$

for the radius of curvature of a plane curve, where r is the distance of the point P of the curve from an origin O and p is the length of the perpendicular from O on the tangent at P (Fig. 15).

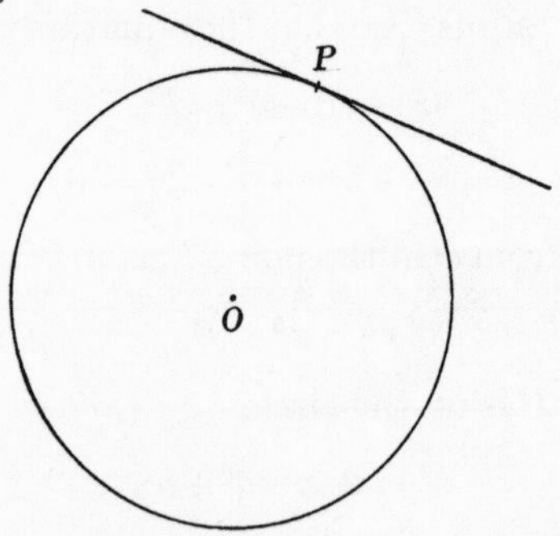

Fig. 15

For a circle, take the origin at the centre, and let the radius (for any one determination of it) be denoted by k. Then $r = p = k$ for all positions of P. Hence p, r are connected by the relation

$$k^m r^n = p^{m+n}$$

for any values of m and n.

Take logarithms:

$$m \log k + n \log r = (m+n) \log p.$$

Differentiate with respect to p:

$$\frac{n}{r} \cdot \frac{dr}{dp} = \frac{m+n}{p},$$

so that

$$r\frac{dr}{dp} = \frac{m+n}{n} \cdot \frac{r^2}{p}$$

$$= \frac{m+n}{n} k$$

$$\therefore \rho = \{1 + (m/n)\} k.$$

Since this is true for any values of m, n, it follows that ρ is indeterminate.

3. The Fallacy that Every Triangle is Isosceles.

Let ABC be an arbitrary triangle. We prove that, necessarily,

$$AB = AC.$$

Denote the lengths of the sides by the usual symbols a, b, c. If D is the middle point of BC (Fig. 16),

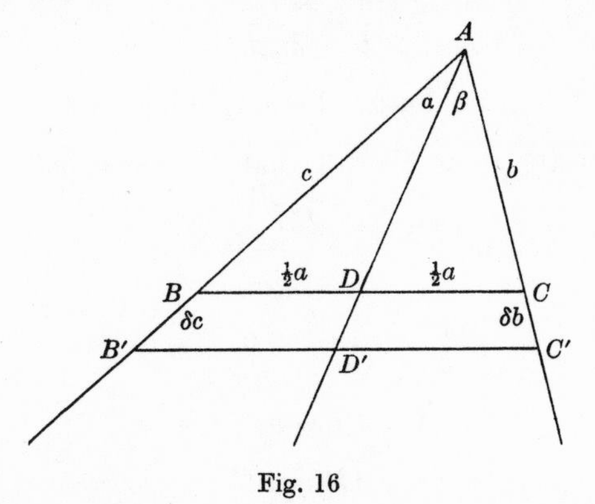

Fig. 16

then, by the sine rule applied to the triangles ADB, ADC,

$$\frac{BD}{\sin BAD} = \frac{AB}{\sin ADB},$$

$$\frac{CD}{\sin CAD} = \frac{AC}{\sin ADC}.$$

Write $\angle BAD = \alpha$, $\angle CAD = \beta$, and divide corresponding sides of these equations, remembering that $BD = CD$; thus

$$\frac{\sin \beta}{\sin \alpha} = \frac{c}{b}.$$

Now subject the base BDC to a small displacement into the position $B'D'C'$ by letting the points B, D, C 'slide', as it were, down AB, AD, AC. The angles α, β are unchanged in this displacement. Adjust the variation, as is legitimate, so that

$$\delta c \equiv BB' = k, \quad \delta b \equiv CC' = k,$$

where the displacement k is small, being the same at both ends of the base. (The line $B'C'$ is not in general parallel to BC for such a displacement.)

Take logarithms of both sides of the relation

$$\frac{c}{b} = \frac{\sin \beta}{\sin \alpha},$$

so that

$$\log c - \log b = \log \sin \beta - \log \sin \alpha.$$

Differentiate, remembering that α, β are constant:

$$\frac{\delta c}{c} - \frac{\delta b}{b} = 0,$$

or

$$\frac{k}{c} - \frac{k}{b} = 0.$$

Hence

$$c = b,$$

so that

$$AB = AC.$$

¶ 4. THE FALLACY THAT $1/x$ IS INDEPENDENT OF x.

We prove first a more general result:

Let x, y, z be given functions of u, v, w. Evaluate $\frac{\partial x}{\partial u}$ and express it in terms of x, y, z. Then $\frac{\partial x}{\partial u}$ so expressed is independent of x.

$\frac{\partial x}{\partial u}$ is a function of x, y, z, and will be independent of x if

$$\frac{\partial}{\partial x}\left(\frac{\partial x}{\partial u}\right) = 0.$$

By definition of second-order partial differentiation, the left-hand side is

$$\frac{\partial^2 x}{\partial x\,\partial u},$$

or, reversing the order of differentiation,

$$\frac{\partial^2 x}{\partial u\,\partial x}.$$

Now this, by definition again, is

$$\frac{\partial}{\partial u}\left(\frac{\partial x}{\partial x}\right);$$

and

$$\frac{\partial x}{\partial x} = 1,$$

so that

$$\frac{\partial}{\partial u}\left(\frac{\partial x}{\partial x}\right) = 0.$$

Hence

$$\frac{\partial}{\partial x}\left(\frac{\partial x}{\partial u}\right) = 0,$$

and so $\frac{\partial x}{\partial u}$ is independent of x.

The particular example quoted above is a corollary for the simple case:

$$x = (2u)^{\frac{1}{2}}, \quad y = (2v)^{\frac{1}{2}}, \quad z = (2w)^{\frac{1}{2}}.$$

Then
$$\frac{\partial x}{\partial u} = (2u)^{-\frac{1}{2}}$$
$$= x^{-1}.$$

Since $\dfrac{\partial x}{\partial u}$ is independent of x, it follows that $1/x$ is independent of x.

II. The Commentary

Fallacy 1.

The difficulty may best be explained by using the language of differentials. The relation
$$r^2 = 2ax - a^2 + b^2$$
is perfectly correct, so that, taking differentials on both sides,
$$r\,dr = a\,dx.$$
For a maximum or minimum value of r, dr must be zero, and this happens (a not being zero) precisely where
$$dx = 0.$$
The turning values of r arrive *at the turning values of* x.

This example emphasises the rule that the independent variable must be unrestricted near the turning values of the dependent variable. We might, for example, have avoided the danger by choosing polar coordinates instead of Cartesian, giving the relation (with $x = r\cos\theta$)
$$r^2 = 2ar\cos\theta - a^2 + b^2.$$
The equation in differentials is now
$$r\,dr = a\cos\theta\,dr - ar\sin\theta\,d\theta,$$
so that dr is zero when
$$ar\sin\theta = 0.$$

Since r cannot be zero (unless $a^2 = b^2$), the condition for a turning value is

$$\sin\theta = 0,$$

so that

$$\theta = 0, \quad \theta = \pi.$$

This gives the two ends of the diameter through O.

FALLACY 2.

This is akin to the preceding fallacy. The relation

$$m\log k + n\log r = (m+n)\log p$$

is correct, and the relation in differentials is

$$\frac{n\,dr}{r} = \frac{(m+n)\,dp}{p}.$$

This is perfectly true. But since r and p are both constant for the circle, dr and dp are both zero, so that no further deduction can be drawn.

The real interest of this fallacy, however, lies in a more detailed examination of the equation

$$k^m r^n = p^{m+n}.$$

To make ideas precise, take the simplest case, when $m = n = 1$, so that the curve is

$$p^2 = kr.$$

This we recognise as the 'pedal' or 'p–r' equation of a *parabola* whose *latus rectum* is $4k$ and whose focus is at the origin. To verify this statement geometrically, let P be an arbitrary point on the parabola of focus S and vertex A, where $SA = k$ (Fig. 17). If the tangent at P meets the tangent at the vertex at T, it is known that

(i) ST is perpendicular to the tangent,

(ii) $\angle AST = \angle TSP$.

Hence the triangles AST, TSP are similar, so that

$$\frac{AS}{ST} = \frac{TS}{SP},$$

or, in the notation of the text,

$$\frac{k}{p} = \frac{p}{r},$$

so that

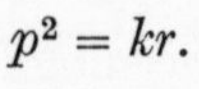

$$p^2 = kr.$$

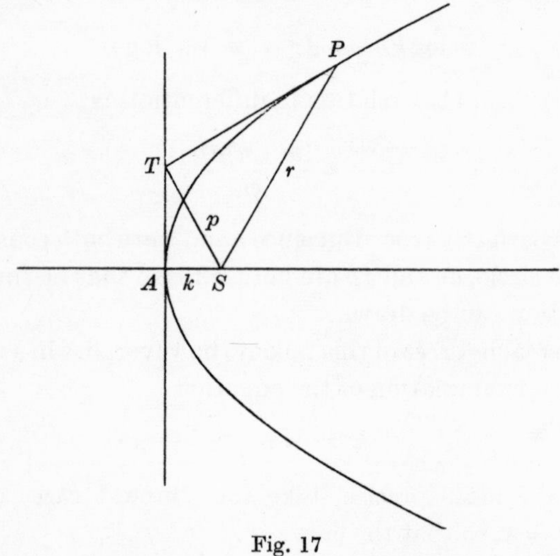

Fig. 17

Hence *the equation*

$$p^2 = kr$$

not only represents (for $p = r = k$*) the circle of centre the origin and radius* k*, but also a parabola of focus the origin and latus rectum* $4k$. The analysis of the text shows that at a general point of the parabola the radius of curvature is $2r^2/p$.

More generally, an equation between r and p of the form

$$f(r, p) = 0$$

gives a certain curve (as in standard theory); but *it is also satisfied by the circles whose radii are the roots of the equation*

$$f(r, r) = 0.$$

These are the circles through the *apses* of the curve, that is, the points where the tangent is perpendicular to the radius. (The formula $\rho = r\,dr/dp$ requires care at such points of the curve, since dr and dp are both zero. See also the preceding fallacy.)

FALLACY 3.

The explanation of this fallacy is comparatively simple, for it is a case of pure misconduct. The formula

$$\frac{c}{b} = \frac{\sin\beta}{\sin\alpha}$$

requires D to be the middle point of BC; but D' is not the middle point of $B'C'$, so the differentiation is not legitimate—α and β would have to vary too.

It is essential to check that variations do not violate the conditions on which formulae are based.

FALLACY 4.

This is an example of hypnosis; the notation

$$\frac{\partial^2 x}{\partial x\,\partial u}, \quad \frac{\partial^2 x}{\partial u\,\partial x}$$

lures one into believing that the differentiations can be reversed in order. But

$$\frac{\partial}{\partial x}\left(\frac{\partial x}{\partial u}\right)$$

is formed on the understanding that the other coefficients are

$$\frac{\partial}{\partial y}\left(\frac{\partial x}{\partial u}\right), \quad \frac{\partial}{\partial z}\left(\frac{\partial x}{\partial u}\right),$$

whereas
$$\frac{\partial}{\partial u}\left(\frac{\partial x}{\partial x}\right)$$

implies
$$\frac{\partial}{\partial v}\left(\frac{\partial x}{\partial x}\right), \quad \frac{\partial}{w}\left(\frac{\partial x}{\partial x}\right).$$

The symbol
$$\frac{\partial}{\partial x}\left(\frac{\partial x}{\partial u}\right)$$

is short for
$$\underset{(y,\, z \text{ constant})}{\frac{\partial}{\partial x}} \quad \underset{(v,\, w \text{ constant})}{\left\{\frac{\partial x}{\partial u}\right\}}$$

and the abbreviation to
$$\frac{\partial^2 x}{\partial x\, \partial u}$$

is unwarranted.

CHAPTER VIII

FALLACIES IN INTEGRATION

The fallacies arise chiefly through failure to observe certain well-known elementary precautions. The results to which this failure leads may emphasise more vividly than warnings the need for care.

I. The Fallacies

1. The Fallacy that 0 = 1.

Consider the integral

$$I \equiv \int \frac{dx}{x}.$$

Integrate by parts:

$$\begin{aligned} I &= \int 1.(1/x)\,dx \\ &= x(1/x) - \int x(-1/x^2)\,dx \\ &= 1 + \int \frac{dx}{x} \\ &= 1 + I. \end{aligned}$$

Hence $\qquad 0 = 1.$

2. The Fallacy that 2 = 1.

Let $f(x)$ be any given function. Then

$$\int_1^2 f(x)\,dx = \int_0^2 f(x)\,dx - \int_0^1 f(x)\,dx.$$

If we write $x = 2y$ in the first integral on the right, then

$$\begin{aligned} \int_0^2 f(x)\,dx &= 2\int_0^1 f(2y)\,dy \\ &= 2\int_0^1 f(2x)\,dx, \end{aligned}$$

on renaming the variable.

Suppose, in particular, that the function $f(x)$ is such that

$$f(2x) \equiv \tfrac{1}{2}f(x)$$

for all values of x. Then

$$\int_1^2 f(x)\,dx = 2\int_0^1 \tfrac{1}{2}f(x)\,dx - \int_0^1 f(x)\,dx$$

$$= 0.$$

Now the relation

$$f(2x) \equiv \tfrac{1}{2}f(x)$$

is satisfied by the function

$$f(x) \equiv \frac{1}{x}.$$

Hence $$\int_1^2 \frac{dx}{x} = 0,$$

so that $$\log 2 = 0,$$

or $$2 = 1.$$

¶ 3. The Fallacy that $\pi = 0$.

To prove that, if $f(\theta)$ is any function of θ, then

$$\int_0^\pi f(\theta)\cos\theta\,d\theta = 0.$$

Substitute $$\sin\theta = t$$

so that $$\cos\theta\,d\theta = dt,$$

and write $$f\{\sin^{-1}t\} \equiv g(t).$$

The limits of integration are 0, 0, since $\sin 0 = 0$ and $\sin\pi = 0$. Hence the integral is

$$\int_0^0 g(t)\,dt = 0.$$

Corollary. The special case when $f(\theta) \equiv \cos\theta$ is of interest. Then the integral is

$$\int_0^\pi \cos^2\theta\, d\theta = \tfrac{1}{2}\int_0^\pi (1+\cos 2\theta)\, d\theta$$

$$= \left[\tfrac{1}{2}\theta + \tfrac{1}{4}\sin 2\theta\right]_0^\pi$$

$$= \tfrac{1}{2}\pi.$$

Hence $$\tfrac{1}{2}\pi = 0.$$

¶ 4. The Fallacy that $\pi = 0$.

Consider the integral

$$I \equiv \int_{2\cos A}^{1} \frac{a\,dx}{\sqrt{(1-a^2x^2)}}$$

$$= \left[\sin^{-1}(ax)\right]_{2\cos A}^{1}$$

$$= \sin^{-1} a - \sin^{-1}(2a\cos A).$$

Suppose, in particular, that $a = \sin A$. Then

$$I = \sin^{-1}(\sin A) - \sin^{-1}(\sin 2A)$$

$$= A - 2A = -A.$$

Now let $$A = \tfrac{1}{3}\pi$$

so that $$2\cos A = 1.$$

Then $$I = \int_1^1 \frac{a\,dx}{\sqrt{(1-a^2x^2)}}$$

$$= 0.$$

Thus $$0 = -A$$

$$= -\tfrac{1}{3}\pi,$$

and so $$\pi = 0.$$

❡ 5. THE FALLACY THAT $\pi = 2\sqrt{2}$.

Consider the integral

$$I \equiv \int_0^\pi x f(\sin x)\, dx,$$

where $f(\sin x)$ is any function of $\sin x$. In accordance with a standard treatment, make the substitution

$$x = \pi - x',$$

and then drop dashes. Thus

$$I = \int_\pi^0 (\pi - x)\, f\{\sin(\pi - x)\}\, d(-x)$$

$$= \int_0^\pi (\pi - x)\, f(\sin x)\, dx.$$

Hence $$2\int_0^\pi x f(\sin x)\, dx = \pi \int_0^\pi f(\sin x)\, dx.$$

Take, in particular,

$$f(u) \equiv u \sin^{-1}(u),$$

so that $$f(\sin x) = \sin x \,.\, x = x \sin x.$$

Then the relation is

$$2\int_0^\pi x^2 \sin x\, dx = \pi \int_0^\pi x \sin x\, dx.$$

But $$\int_0^\pi x^2 \sin x\, dx = \pi^2 - 4,$$

$$\int_0^\pi x \sin x\, dx = \pi.$$

Hence $$2(\pi^2 - 4) = \pi^2,$$

or $$\pi^2 = 8,$$

so that $$\pi = 2\sqrt{2}.$$

¶ 6. The Fallacy that a Cycloid has Arches of Zero Length.

To prove that the length of an arch of the curve

$$x = t + \sin t, \quad y = 1 + \cos t$$

is zero.

An arch is covered by values of t running from 0 to 2π. If s denotes length of arc, then the required length is, by standard formula,

$$\int_0^{2\pi} \frac{ds}{dt}\, dt.$$

Now
$$\frac{dx}{dt} = 1 + \cos t, \quad \frac{dy}{dt} = -\sin t,$$

so that
$$\begin{aligned}\left(\frac{ds}{dt}\right)^2 &= (1 + \cos t)^2 + (-\sin t)^2 \\ &= 1 + 2\cos t + \cos^2 t + \sin^2 t \\ &= 2 + 2\cos t \\ &= 2(1 + \cos t) \\ &= 4\cos^2 \tfrac{1}{2}t.\end{aligned}$$

Thus the length is

$$\begin{aligned}\int_0^{2\pi} 2\cos \tfrac{1}{2}t\, dt &= \Big[4 \sin \tfrac{1}{2}t\Big]_0^{2\pi} \\ &= 0.\end{aligned}$$

II. Commentary

Fallacy 1.

The error is brought about through the omission of the arbitrary constant for indefinite integrals. The anomaly disappears for definite integration:

If

$$I \equiv \int_a^b \frac{dx}{x},$$

then $$I = \Big[x(1/x)\Big]_a^b - \int_a^b x(-1/x^2)\,dx$$

$$= \Big[1\Big]_a^b + \int_a^b \frac{dx}{x}$$

$$= 0 + I$$

$$= I.$$

(Note that $\Big[1\Big]_a^b = 0$ since the function 1 has the value 1 at $x = b$ and also at $x = a$.)

FALLACY 2.

The integral

$$\int_0^1 f(x)\,dx$$

does not exist when $f(x) \equiv 1/x$. A reconciliation statement may be effected by returning to the original integral. It is true that for any value of δ greater than zero

$$\int_1^2 \frac{dx}{x} = \int_\delta^2 \frac{dx}{x} - \int_\delta^1 \frac{dx}{x}.$$

Put $x = 2y$ in the first integral on the right; it becomes

$$\int_{\frac{1}{2}\delta}^1 \frac{2\,dx}{2x} = \int_{\frac{1}{2}\delta}^1 \frac{dx}{x}.$$

The relation is thus

$$\int_1^2 \frac{dx}{x} = \int_{\frac{1}{2}\delta}^1 \frac{dx}{x} - \int_\delta^1 \frac{dx}{x},$$

$$= \int_{\frac{1}{2}\delta}^\delta \frac{dx}{x}.$$

The assumption made in the text is that this integral tends to zero with δ—as is indeed plausible, since both limits then vanish. But we know that

$$\int_{\frac{1}{2}\delta}^{\delta} \frac{dx}{x} = \Big[\log x\Big]_{\frac{1}{2}\delta}^{\delta}$$

$$= \log\delta - \log\tfrac{1}{2}\delta$$

$$= \log\left(\frac{\delta}{\frac{1}{2}\delta}\right)$$

$$= \log 2.$$

This is in fact the value of

$$\int_1^2 \frac{dx}{x},$$

so the problem is resolved.

FALLACY 3.

The fallacy may be exposed by considering the particular example in detail. Let

$$I \equiv \int_0^{\pi} \cos^2\theta\, d\theta.$$

Substitute $$\sin\theta = t,$$

so that $$\cos\theta\, d\theta = dt.$$

Then, as an intermediate step,

$$I = \int \cos\theta\, dt$$

between appropriate limits.

Now $\cos\theta$ is given in terms of t by the relation

$$\cos\theta = \pm\sqrt{(1-t^2)},$$

where the positive sign must be taken for the part $(0, \frac{1}{2}\pi)$ *of the interval of integration and the negative sign for the part* $(\frac{1}{2}\pi, \pi)$. Thus

$$I = +\int \sqrt{(1-t^2)}\,dt - \int \sqrt{(1-t^2)}\,dt,$$

where the first integration corresponds to $0 \leqslant \theta \leqslant \frac{1}{2}\pi$ and the second to $\frac{1}{2}\pi \leqslant \theta \leqslant \pi$, so that the limits for t are 0, 1 and 1, 0, respectively. Hence

$$\begin{aligned} I &= \int_0^1 \sqrt{(1-t^2)}\,dt - \int_1^0 \sqrt{(1-t^2)}\,dt \\ &= 2\int_0^1 \sqrt{(1-t^2)}\,dt \\ &= \tfrac{1}{2}\pi. \end{aligned}$$

FALLACY 4.

It is immediately obvious that the error must lie in the interpretation of the inverse sines in the step

$$\sin^{-1}(\sin A) - \sin^{-1}(\sin 2A) = A - 2A,$$

but it is less easy to see exactly what is correct. We have to examine the expression

$$\Big[\sin^{-1}(x \sin A)\Big]_{2\cos A}^{1},$$

where, to avoid further complications, we suppose that the lower limit is not greater than the upper, so that

$$\cos A \leqslant \tfrac{1}{2},$$

or, for an acute angle,

$$A \geqslant \tfrac{1}{3}\pi.$$

The essential point is that the inverse sine must vary *continuously* as x rises from $2\cos A$ to 1.

Draw the curve

$$v = \frac{\sin u}{\sin A}$$

for values of u between 0 and π (Fig. 18). Since, by the assumption, $A \geqslant \frac{1}{3}\pi$, it follows that $2A \geqslant \frac{2}{3}\pi$, as indicated in the diagram. The values of v for $u = A$, $2A$ are 1, $2\cos A$, respectively, where $2\cos A \leqslant 1$. The line $v = 1$ cuts the curve at P, where $u = A$, and at R, where $u = \pi - A$; the line $v = 2\cos A$ cuts the curve at Q, where $u = 2A$, and at S, where $u = \pi - 2A$.

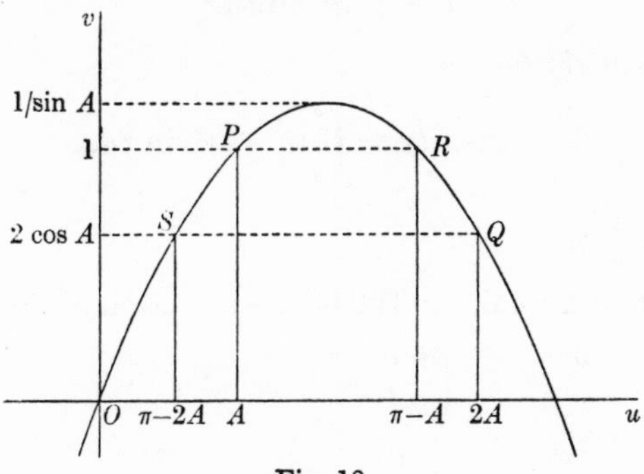

Fig. 18

To get the variation in the inverse sine as v *rises* from $2\cos A$ to 1, it is possible to take *either* the journey from S to P, giving

$$A - (\pi - 2A) = 3A - \pi,$$

or the journey from Q to R, giving

$$(\pi - A) - 2A = \pi - 3A,$$

that choice of sign being taken here which, by the nature of the integral, gives a positive result: that is, $3A - \pi$. Thus

$$I = 3A - \pi,$$

which *vanishes* when $A = \frac{1}{3}\pi$.

The argument on p. 67 wrongly associated P with Q, leading to the fallacy.

FALLACY 5.

This is very similar to its predecessor and need not be given in much detail. The point is that

$$\int_0^\pi (\pi - x) f\{\sin(\pi - x)\}\, dx$$

$$= \int_0^\pi (\pi - x).[\sin x.(\pi - x)]\, dx$$

$$= \int_0^\pi (\pi - x)^2 \sin x\, dx.$$

Since $$I = \int_0^\pi x^2 \sin x\, dx,$$

the correct relation is

$$\int_0^\pi x^2 \sin x\, dx = \int_0^\pi (\pi - x)^2 \sin x\, dx.$$

FALLACY 6.

The error here lies in the false extraction of the square root:

$$\left(\frac{ds}{dt}\right)^2 = 4\cos^2 \tfrac{1}{2}t$$

$$\therefore\ \frac{ds}{dt} = 2\cos \tfrac{1}{2}t.$$

The correct step is $$\frac{ds}{dt} = 2\cos \tfrac{1}{2}t$$

for the interval $(0, \pi)$, and

$$\frac{ds}{dt} = -2\cos \tfrac{1}{2}t$$

for the interval $(\pi, 2\pi)$ where the cosine is negative. Thus

$$S = \int_0^\pi 2\cos \tfrac{1}{2}t\, dt - \int_\pi^{2\pi} 2\cos \tfrac{1}{2}t\, dt$$

$$= \Big[4\sin \tfrac{1}{2}t\Big]_0^\pi - \Big[4\sin \tfrac{1}{2}t\Big]_\pi^{2\pi}$$

$$= 4 - (-4)$$

$$= 8.$$

CHAPTER IX

FALLACY BY THE CIRCULAR POINTS AT INFINITY

I. THE FALLACIES

¶ 1. THE FALLACY THAT THE FOUR POINTS OF INTERSECTION OF TWO CONICS ARE COLLINEAR.

GIVEN: Two conics and a pair of common tangents PU and QV touching one conic at P, Q and the other at U, V. The conics meet in four distinct points A, B, C, D, and the chords of contact PQ, UV meet at R (Fig. 19).

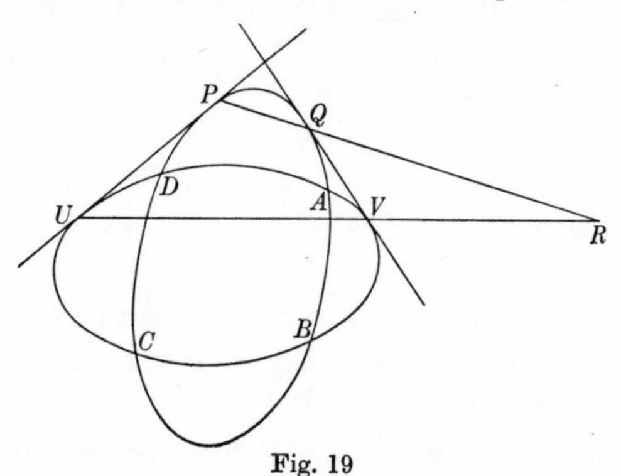

Fig. 19

REQUIRED: To prove that the line DR contains each of the three points A, B, C.

CONSTRUCTION: To prove that, say, A lies on DR, project B, C into the circular points at infinity. The two conics then become *circles*, with A, D as common points and with PU, QV as common tangents (Fig. 20).

PROOF: By symmetry, or by easy independent proof, the lines PQ, UV, AD are all parallel. That is, PQ meets

UV in a point R at infinity, and DA also passes through R. Thus A lies on DR in the projected figure, and so A lies on DR in the original figure.

In the same way, B and C also lie on DR, and so the four points A, B, C, D are collinear.

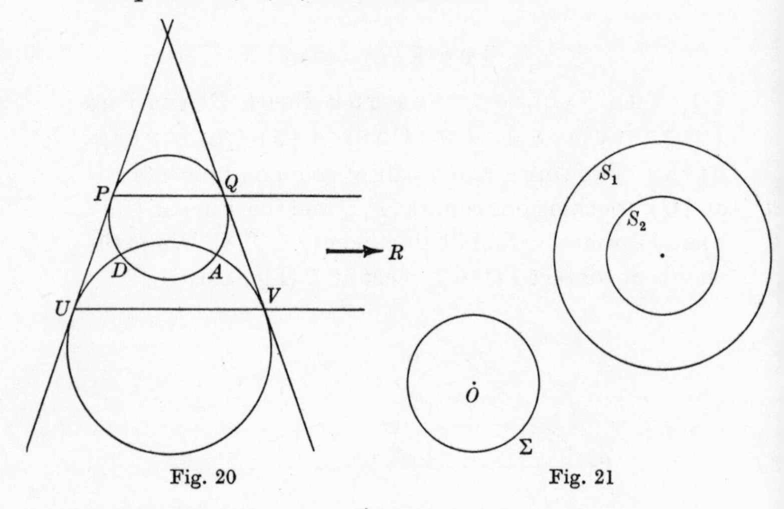

Fig. 20

Fig. 21

§ 2. The Fallacy that Concentric Circles Invert into Concentric Circles with Respect to an Arbitrary Point.

Let S_1, S_2 be two given concentric circles and O an arbitrary point, the centre of a circle Σ (Fig. 21). Invert S_1 and S_2 with respect to Σ.

The circular points I, J lie on Σ, and so each inverts into itself. Moreover, concentric circles *touch* at I and at J, and the property of touching is not affected by inversion. Hence S_1, S_2 invert into cirlces S'_1, S'_2 which also touch at I, J. They are therefore concentric circles.

II. The Commentary

Fallacy 1.

This example illustrates the danger, often ignored, of 'circular points' argument when real and complex geometry are confused. The given theorem belongs to *complex geometry*, where two conics with four distinct points *always* have four common tangents. Projection transfers attention to *real geometry*, where two circles through two points have only two common tangents.

Call the four tangents of the complex geometry $\alpha, \beta, \gamma, \delta$. When B, C are projected into the circular points at infinity, two of them, say β, γ, are projected into the two real tangents of the circles, and it is from them that the collinearity of D, R, A is established. But when two other points, say C, A, are projected into the circular points, it is a different pair of tangents that 'appears' in the real plane, so that the intersection of chords of contact is not R but some other point R'; and all that we have proved is that D, B and this point R' are collinear. Similarly projection of A, B into the circular points gives the collinearity of D, C with yet another point R''.

The basis of the proof is therefore unsound.

Fallacy 2.

There are two things that must be kept in mind about inversion as it is normally understood: first, that the definitions and proofs are based on *real* geometry, and do not necessarily apply to complex; second, that when extensions are made beyond Euclidean geometry the correspondence does not remain (1, 1) for all points of the plane.

Suppose that P, P' are inverse points with respect to the circle Σ of centre O and radius a. Then

$$OP.OP' = a^2,$$

so that
$$OP' = \frac{a^2}{OP}.$$

If P is conceived as moving further and further away from O along a given radius, then P' moves nearer and nearer to O along that radius. In a geometry in which the *line at infinity* is admitted, the 'points' of that line correspond to the directions of approach to O. The 'line' itself is concentrated by inversion into the point O. Since I, J lie on the line at infinity, the language 'each inverts into itself' cannot be accepted without reservation.

But this is not the whole story—otherwise, indeed, we should reach the conclusion that every circle passed through O. The fact that the circular points belong to a complex geometry must now be considered. In homogeneous coordinates, these are the points

$$I(1, i, 0), \quad J(1, -i, 0),$$

each lying on every circle

$$x^2+y^2+2gxz+2fyz+cz^2 = 0.$$

In so far as they can be expressed at all in non-homogeneous coordinates, they are the points lying 'at infinity' along the line

$$\frac{x}{1} = \frac{y}{\pm i},$$

and so the non-homogeneous coordinates of a point P on OI (where the origin is at O) can be expressed in the form

$(\lambda, i\lambda)$. But *such a point is at zero distance from O*, since that distance d is given by the formula

$$\begin{aligned} d^2 &= (x_1 - x_2)^2 + (y_1 - y_2)^2 \\ &= (\lambda - 0)^2 + (i\lambda - 0)^2 \\ &= \lambda^2(1 + i^2) \\ &= 0. \end{aligned}$$

Thus if we allow any meaning to distance in general, and to infinite distance in particular, we come to the conclusion that *the inverse of every point of the line OI is at I.*

We are, in fact, attempting to combine real and complex geometries, and have reached a sensitive spot.

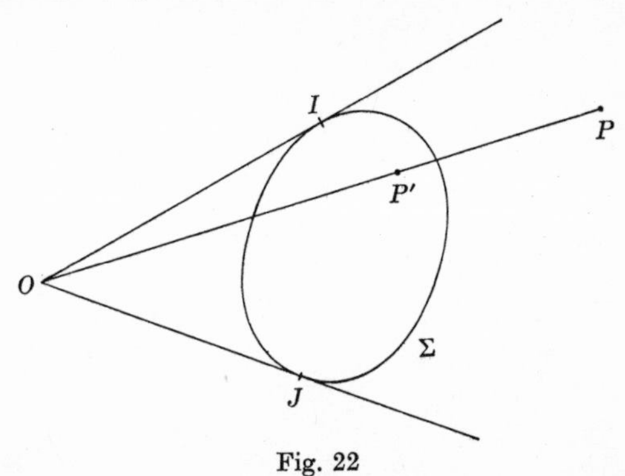

Fig. 22

There is a more satisfactory interpretation in terms of pure geometry. Let the tangents at points I, J on a given conic Σ meet at O (Fig. 22). From any point P in the plane another point P' can be defined as follows:

Join OP, and let P' be the point on OP such that P, P' are conjugate with respect to Σ, that is, such that the polar of either passes through the other.

This defines P' uniquely for given P, and, when I, J are the circular points at infinity, the two points are inverse with respect to Σ.

The cases of exception are readily seen:

(*a*) If P lies on IJ, its polar passes through O and so meets OP at O. Thus *all points of the line IJ have their inverses at O.*

(*b*) If P is on OI, then the polar of P passes through I and therefore meets OP at I. Thus *all points of the line OI* (*or OJ*) *have their inverses* at I (*or J*)

We have therefore broken down the statement (p. 76) that I inverts into itself; its inverse is not unique, but consists of every point on the line OI.

CHAPTER X

SOME 'LIMIT' FALLACIES

I. The Fallacies

¶ 1. The Binomial Fallacy.

The binomial theorem states that

$$(a+b)^n = a^n + na^{n-1}b + \frac{n(n-1)}{2!}a^{n-2}b^2 + \ldots + nab^{n-1} + b^n.$$

Put $n = 0$. Then

$$(a+b)^0 = 1, \quad a^0 = 1, \quad b^0 = 1;$$

and all terms on the right except the two outside ones have n as a factor and therefore vanish. Thus

$$\begin{aligned} 1 &= 1+0+0+\ldots+0+1 \\ &= 2. \end{aligned}$$

¶ 2. The Fallacy that all Numbers are Equal.

Let m, n be two given numbers, and consider the function

$$\frac{mx+ny}{x+y}.$$

Whenever y is 0, its value is

$$\frac{mx}{x},$$

or

$$m,$$

and whenever x is 0, its value is

$$\frac{ny}{y},$$

or

$$n.$$

Included in each of these two cases is the particular case when $x = y = 0$. Since $y = 0$, the value of the function is m; since $x = 0$, the value of the function is n. Hence

$$m = n.$$

(The reader suspicious of the step

$$\frac{mx}{x} = m$$

at $x = 0$ should note that, in any case,

$$\lim_{x \to 0} \frac{mx}{x} = m$$

and, similarly, that

$$\lim_{y \to 0} \frac{ny}{y} = n.)$$

❡ 3. The Further Fallacy that all Numbers are Equal.

Let m, n be any two numbers, and let x denote their difference, so that

$$x = m - n.$$

Then $$m\{x^2 - (m-n)\,x\} = n\{x^2 - (m-n)\,x\}.$$

It is, of course, illegitimate to divide this equation by $x^2 - (m-n)\,x$, which is zero, but, alternatively, we may use the limiting argument

$$\frac{m}{n} = \lim_{x \to m-n} \frac{x^2 - (m-n)\,x}{x^2 - (m-n)\,x}.$$

Now we know that, if $f(x)$, $g(x)$ are two functions such that $f(k) = 0$, $g(k) = 0$, but such that $g'(k) \neq 0$, where $g'(k)$ is the differential coefficient, then

$$\lim_{x \to k} \frac{f(x)}{g(x)} = \frac{f'(k)}{g'(k)}.$$

Here, $k = m - n$ and

$$f(x) = g(x) = x^2 - (m-n)x,$$

so that $$f'(x) = g'(x) = 2x - (m-n)$$

and $$f'(m-n) = g'(m-n) = m-n.$$

If $m = n$, there is nothing to prove. If $m \neq n$, then $m - n \neq 0$, so that $g'(m-n) \neq 0$. Hence

$$\lim_{x \to m-n} \frac{x^2-(m-n)x}{x^2-(m-n)x} = \frac{m-n}{m-n} = 1.$$

Thus $$\frac{m}{n} = 1,$$

or $$m = n.$$

4. Some Infinite Series Fallacies.

A number of somewhat synthetic fallacies (some of them, though, important in their day) can be obtained from infinite series. The following examples are reasonably brief.

(i) TO PROVE THAT 1 = 0. Write

$$S \equiv 1-1+1-1+1-1+\ldots.$$

Then, grouping in pairs,

$$\begin{aligned} S &= (1-1)+(1-1)+(1-1)+\ldots \\ &= 0+0+0+\ldots \\ &= 0. \end{aligned}$$

Also, grouping alternatively in pairs,

$$\begin{aligned} S &= 1-(1-1)-(1-1)-(1-1)-\ldots \\ &= 1-0-0-0-\ldots \\ &= 1. \end{aligned}$$

Hence $$1 = 0.$$

(ii) TO PROVE THAT -1 IS POSITIVE. Write

$$S \equiv 1+2+4+8+16+32+\ldots.$$

Then S is positive. Also, multiplying each side by 2,

$$2S = 2+4+8+16+32+\ldots$$
$$= S-1.$$

Hence $$S = -1,$$

so that -1 is positive.

(iii) TO PROVE THAT 0 IS POSITIVE (THAT IS, GREATER THAN ZERO).

Write $$u = 1+\tfrac{1}{3}+\tfrac{1}{5}+\tfrac{1}{7}+\ldots,$$
$$v = \tfrac{1}{2}+\tfrac{1}{4}+\tfrac{1}{6}+\tfrac{1}{8}+\ldots.$$

Then $$2v = 1+\tfrac{1}{2}+\tfrac{1}{3}+\tfrac{1}{4}+\ldots$$
$$= u+v,$$

so that $$u-v = 0.$$

But, on subtracting corresponding terms,

$$u-v = (1-\tfrac{1}{2})+(\tfrac{1}{3}-\tfrac{1}{4})+(\tfrac{1}{5}-\tfrac{1}{6})+(\tfrac{1}{7}-\tfrac{1}{8})+\ldots,$$

where *each* bracketed term is *greater* than zero. Thus

$u-v$ is greater than zero,

or 0 is greater than zero.

II. The Commentary

FALLACY 1.

It is easy to dispose of the fallacy in this argument, for the binomial theorem for positive integers begins at $n = 1$ and the proof does not apply when $n = 0$. It is less easy to see where the extra 1 comes from, and the argument is worth examining in detail.

Since the binomial expansion, when n is not a positive integer, involves an infinite series, we take, in accordance with normal practice, the value $a = 1$ and assume that $|b| < 1$. Consider, then, the series

$$1 + nb + \frac{n(n-1)b^2}{2!} + \frac{n(n-1)(n-2)b^3}{3!} + \ldots,$$

in which the coefficient of b^k, where k is a positive integer, is

$$\frac{n(n-1)(n-2)\ldots(n-k+1)}{k(k-1)(k-2)\ldots 1}.$$

This is a function of n and k, which we now examine on its own merits, ignoring the restriction for n to be a positive integer. Write

$$f(n,k) \equiv \frac{n(n-1)(n-2)\ldots(n-k+1)}{k(k-1)(k-2)\ldots 1}.$$

Let n approach the value k; thus

$$\lim_{n\to k} f(n,k) = 1.$$

Now let k approach the value 0; thus

$$\lim_{k\to 0}\left\{\lim_{n\to k} f(n,k)\right\} \equiv \lim_{k\to 0}(1) = 1.$$

Alternatively, let n approach the value 0; thus

$$\lim_{n\to 0} f(n,k) = 0.$$

Now let k approach the value 0; thus

$$\lim_{k\to 0}\left\{\lim_{n\to 0} f(n,k)\right\} \equiv \lim_{k\to 0}(0) = 0.$$

Both limiting processes finish with n and k zero, but the different approach gives a result which is 1 in the first case and 0 in the second.

These results may be applied to the particular fallacy.

For the first, if n is identified with the positive integer k, then a coefficient of value 1 is obtained. The *subsequent* step $k = 0$ is meaningless as picking out a term of the series, but the value 1 persists arithmetically, giving (since $b^0 = 1$) an extra 1 in the expansion.

In the second case, however, the immediate identification of n with zero precludes the possibility of an additional term, and so leads to the correct result.

This example is very instructive as affording a simple illustration of the way in which the passage to limiting values in two stages may be profoundly affected by the order in which the calculations are performed.

FALLACY 2.

This fallacy, like the preceding one, gives an illustration of the care that must be taken when two limiting processes are reversed in order of operation.

A slight change of notation allows the use of the language of polar coordinates. Consider the function

$$u \equiv \frac{mx^2 + ny^2}{x^2 + y^2}.$$

In terms of polar coordinates, with

$$x = r\cos\theta, \quad y = r\sin\theta$$

(so that $x^2 + y^2 = r^2$), we have

$$u = m\cos^2\theta + n\sin^2\theta.$$

The dénouement $\quad x = 0, \quad y = 0$

arises when $r = 0$, but the limiting processes towards that end depend on θ as well as r. The origin $x = 0$, $y = 0$ may be regarded as approached along a line $\theta = \alpha$, and the limit is then

$$m\cos^2\alpha + n\sin^2\alpha.$$

But *this depends on* α. In particular, it has the value m when $\alpha = 0$ and the value n when $\alpha = \frac{1}{2}\pi$, corresponding to the two particular limits given in the text.

FALLACY 3.

The fallacy itself is rather elementary, but it does serve to emphasise a point that is sometimes ignored. The trouble lies in the expression

$$\lim_{x \to m-n} \frac{x^2 - (m-n)x}{x^2 - (m-n)x},$$

where the limiting process itself is meaningless, since x always has precisely the value $m-n$ and never any other. The subsequent theory therefore collapses.

FALLACY 4.

These fallacies are included chiefly because the collection might appear incomplete without some reference to infinite series. They are brought about by ignoring standard rules governing the convergence of series, details of which may be found in appropriate text-books. The series are all non-convergent.

There is some interest in a comparison of (ii) with the corresponding correct form for a convergent series:

Write $$S \equiv 1 + \tfrac{1}{2} + \tfrac{1}{4} + \tfrac{1}{8} + \ldots,$$

so that $$2S = 2 + 1 + \tfrac{1}{2} + \tfrac{1}{4} + \ldots$$

$$= 2 + S.$$

Hence $$S = 2.$$

CHAPTER XI

SOME MISCELLANEOUS HOWLERS

It is believed that the howlers with which we conclude this book are all genuine, in the sense that they were perpetrated innocently in the course of class study or of examination. Teachers will be familiar with the type of mind producing them, and little comment seems necessary.

The first two howlers are so astonishing that an investigation is added giving the generalised theory under which they become possible. To get the full benefit from the others, the reader should trace the mistakes to their source, not being content with merely locating the wrong step: for example, the statement

$$36^2 = 336$$

in no. 11 is the stroke of genius on which the whole solution turns, and arises, presumably, from the argument

$$36^2 = 3(6)^2 = 3(36) = 336.$$

The author cannot follow the subsequent step

$$x^2 + x^2 - x^2 = x^4.$$

1. *To solve the equation*

$$(x+3)(2-x) = 4.$$

Either $\qquad x+3 = 4 \quad \therefore x = 1,$

or $\qquad 2-x = 4 \quad \therefore x = -2.$

Correct.

Commentary. Every quadratic equation can be reduced to a form leading to this method of solution:

The quadratic equation with roots p, q is

$$(x-p)(x-q) = 0$$

$$\therefore\ x^2-px-qx+pq = 0$$

$$\therefore\ 1-p+q = 1+q-x-p-pq+px+x+qx-x^2$$

$$= (1+q-x)-p(1+q-x)+x(1+q-x)$$

$$= (1+q-x)(1-p+x)$$

$$\therefore\ (1+q-x)(1-p+x) = 1-p+q$$

'$\therefore$' ***either*** $\quad 1+q-x = 1-p+q$

$$\therefore\ x = p,$$

or $\quad 1-p+x = 1-p+q$

$$\therefore\ x = q.$$

2. *To find the largest angle of the triangle with sides* 4, 7, 9.

$$\sin C = \tfrac{9}{7}$$

$$= 1{\cdot}2857.$$

But $\quad 1 = \sin 90°,$

and $\quad {\cdot}2857 = \sin 16°\,36'$

$$\therefore\ 1{\cdot}2857 = \sin 106°\,36'$$

$$\therefore\ C = 106°\,36'.$$

Correct.

Commentary. Let OBC be a triangle in which the angle at C is a right angle. Draw the circle with centre B and radius $BO+BC$ to cut the circle with centre C and radius $CO+CB$ in A. Then ABC is a triangle whose angle C can be found by the above method.

Let $BC = a$, $OC = x$, $OB = y$. Then

$$CA = x+a, \quad AB = y+a$$

and $\quad y^2-x^2 = a^2.$

The method gives

$$\text{`}\sin C = \frac{y+a}{x+a} = 1 + \frac{y-x}{x+a}.$$

If, then,

$$\sin\theta = \frac{y-x}{x+a},$$

$$C = 90^\circ + \theta.\text{'}$$

On this basis, $\cos C = -\sin\theta = \dfrac{x-y}{x+a}$.

But the more usual formula gives

$$\cos C = \frac{a^2+b^2-c^2}{2ab} = \frac{a^2+(x+a)^2-(y+a)^2}{2a(x+a)}$$

$$= \frac{x^2-y^2+a^2+2ax-2ay}{2a(x+a)} = \frac{2ax-2ay}{2a(x+a)} = \frac{x-y}{x+a}.$$

The two formulae therefore agree.

3. *To prove that, if*

$$I_n = \int_0^{\frac{1}{2}\pi} x^n \sin x\,dx,$$

then $$I_n + n(n-1)I_{n-2} = n(\tfrac{1}{2}\pi)^{n-1}.$$

On 'integration by parts',

$$I_n = \int_0^{\frac{1}{2}\pi} x^n \sin x\,dx$$

$$= \Big[nx^{n-1}\sin x\Big]_0^{\frac{1}{2}\pi} - \int_0^{\frac{1}{2}\pi} nx^{n-1}\cos x\,dx$$

$$= n(\tfrac{1}{2}\pi)^{n-1} - \Big[n(n-1)x^{n-2}\cos x\Big]_0^{\frac{1}{2}\pi}$$

$$+ \int_0^{\frac{1}{2}\pi} n(n-1)x^{n-2}(-\sin x)\,dx$$

$$= n(\tfrac{1}{2}\pi)^{n-1} - 0 - n(n-1)I_{n-2}$$

$$\therefore\ I_n + n(n-1)I_{n-2} = n(\tfrac{1}{2}\pi)^{n-1}.$$

4. *To prove the converse of the theorem of Apollonius, that, if D is a point in the base BC of a triangle ABC such that*

$$AB^2+AC^2 = 2DA^2+DB^2+DC^2,$$

then D is the middle point of BC.

Let E be the middle point of BC. Then, by the theorem of Apollonius,

$$AB^2+AC^2 = 2EA^2+2EB^2$$

$$= 2EA^2+EB^2+EC^2.$$

Thus $\quad 2DA^2+DB^2+DC^2 = 2EA^2+EB^2+EC^2$

$$\therefore\ D(2A^2+B^2+C^2) = E(2A^2+B^2+C^2).$$

But $\quad 2A^2+B^2+C^2 \neq 0,$

otherwise A, B, C would all be zero. Hence

$$D = E,$$

so that D is the middle point of AB.

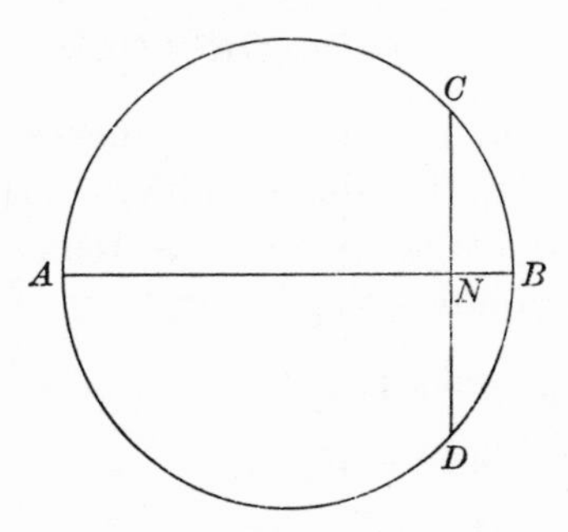

Fig. 23

5. *To prove that, if CD is a chord of a circle perpendicular to a diameter AB and meeting it in N, then the sum of the areas of the circles on AN, BN, CN, DN as diameters is equal to the area of the whole circle* (Fig. 23).

The sum of the areas is

$$\tfrac{1}{4}\pi(AN^2+BN^2+CN^2+DN^2)$$
$$= \tfrac{1}{4}\pi N^2(A+B+C+D).$$

But A, B, C, D are all points on the circumference of the big circle. Hence the sum of the areas is equal to the area of the whole circle.

6. *To prove that, if O is a point inside a rectangle $ABCD$, then $OA^2+OC^2 = OB^2+OD^2$.*

If the equation is divided by O, its value is unchanged. Thus

$$A^2+C^2 = B^2+D^2.$$

Now $$A = B = C = D$$

because the figure is a rectangle. Thus

$$A^2 = B^2 = C^2 = D^2,$$

and so $$A^2+C^2 = B^2+D^2,$$

but this is equal to

$$OA^2+OC^2 = OB^2+OD^2.$$

Commentary. The fusion of the branches algebra, geometry, etc., into the single subject mathematics is strongly urged by many teachers today. The above examples afford interesting illustrations of the process.

7. *To prove that* $8^{-\frac{5}{3}} \times 4^{\frac{5}{2}} = 1$.

(i) $2^{-5} \times 2^5 = 4^{5-5} = 4^0 = 1.$

(ii) $\dfrac{8^3}{8^5} \times \dfrac{4^5}{4^2} = \dfrac{4^3}{8^2} = \dfrac{64}{64} = 1.$

(iii) $(4^{\frac{1}{2}})^5 \div (8^{\frac{1}{3}})^5 = 4^{\frac{1}{2}} \div 8^{\frac{1}{3}} = 2 \div 2 = 1.$

(iv) $5\sqrt{4} \div 5\sqrt[3]{8} = 10 \div 10 = 1.$

(v) $\frac{5}{2}\log 4 \div \frac{5}{3}\log 8 = 5\log 2 \div 5\log 2 = 1.$

8. *To solve the equation* $2\sqrt{x} = x - 3$.

$$2x^{\frac{1}{2}} = x - 3$$
$$\therefore \; (x^{\frac{1}{2}})^{\frac{4}{2}} = x - 3$$
$$\therefore \; x^{\frac{5}{2}} = x - 3$$
$$\therefore \; x^{\frac{5}{2}} - x^{\frac{2}{2}} = 3$$
$$\therefore \; x^3 = 3$$
$$\therefore \; x = 3 \times 3 = 9.$$

Correct.

9. *To simplify the expression* $\dfrac{2^3 . 4^8 . 16^{-n}}{16^2 . 4^{-2n}}$.

$$\frac{22^{11-n}}{20^{2-2n}}$$
$$= 22^{11-n} - 20^n$$
$$= 2^{11}.$$

Correct.

10. *To solve the equation*

$$(5-3x)(7-2x) = (11-6x)(3-x).$$
$$5-3x+7-2x = 11-6x+3-x$$
$$\therefore \; 12-5x = 14-7x$$
$$\therefore \; 2x = 2$$
$$\therefore \; x = 1.$$

Correct.

11. *To solve the equation*

$$x^2+(x+4)^2 = (x+36)^2.$$
$$x^2+x^2+4^2 = x^2+36^2$$
$$\therefore \; x^2+x^2+16 = x^2+336$$
$$\therefore \; x^2+x^2-x^2 = 336-16$$
$$\therefore \; x^4 = 320$$
$$\therefore \; x = 80.$$

Correct.

12. *To prove that the series*

$$\sum_{n=1}^{\infty} \frac{i^n}{n} \qquad (i = \sqrt{(-1)})$$

converges.

Applying the ratio test,

$$\frac{u_{n+1}}{u_n} = \frac{n}{n+1} i$$

$$\to i.$$

But
$$-1 < 1$$

$$\therefore \sqrt{(-1)} < \sqrt{1}$$

$$\therefore i < 1.$$

Hence
$$\lim_{n\to\infty} \frac{u_{n+1}}{u_n} < 1,$$

so that the series converges.

13. *To prove that*

$$1 + 2x + 2x^2 + x^3 \equiv \frac{(1+x)(1-x^3)}{1-x}.$$

Put $x = 1$. Then

$$1 + 2 + 2 + 1 = \frac{2(1-1)}{1-1}$$

$$\therefore 6 = \frac{2 \times 0}{0}.$$

Since
$$\frac{0}{0} = \infty$$

and
$$\text{infinity} = \text{anything},$$

then it is equal.

14. *To prove that, if*

$$\frac{a+b}{b+c} = \frac{c+d}{d+a},$$

then ***either*** $a = c$ ***or*** $a + b + c + d = 0$.

Each ratio is

$$\frac{a+b+c+d}{b+c+d+a}$$

$$= 1.$$

Hence $$a = c.$$

If $a \neq c$, then

$$a+b \neq b+c$$

$$c+d \neq d+a$$

$$\therefore\ a+b+c+d \neq a+b+c+d,$$

which is not true unless

$$a+b+c+d = 0.$$